Science and Technology in Ancient Indian Texts

Science and Technology in Ancient Indian Texts

Editors

Bal Ram Singh

Girish Nath Jha

Umesh Kumar Singh

Diwakar Mishra

PRINTWORLD

Publishers of Indian Traditions

Cataloging in Publication Data — DK
[Courtesy: D.K. Agencies (P) Ltd. <docinfo@dkagencies.com>]

Seminar on Science and Technology in Ancient Indian Texts (2010 : New Delhi, India)
Science and technology in ancient Indian texts / editors, Bal Ram Singh, Girish Nath Jha, Umesh Kumar Singh, Diwakar Mishra.
pages cm
Includes passages in Sanskrit (roman).
"First published in India in 2012" – Title page verso.
Copyright: University of Massachusetts Dartmouth, USA and Jawaharlal Nehru University, New Delhi.
Papers presented at a Seminar on Science and Technology in Ancient Indian Texts, organized at the Special Centre for Sanskrit Studies, Jawaharlal Nehru University on 9-10 January 2010 with support of Center for Indic Studies, University of Masachusetts, Dartmouth, US.
Includes bibliographical references and index.
ISBN 9788124606322 (HB)

1. Vedic literature – History and criticism – Congresses. 2. Vedas and science – Congresses. 3. Science, Ancient – Congresses. I. Singh, Bal Ram, editor. II. Jawaharlal Nehru University. Special Centre for Sanskrit Studies, host institution. III. University of Massachusetts Dartmouth. Center for Indic Studies, organizer. IV. Title.

LCC BL1112.27.S46 2010 | DDC 294.5921046 23

ISBN 978-81-246-0632-2
First Published in India in 2012
Reprinted in 2024

Printed and published by:
D.K. Printworld (P) Ltd.
Regd. Office: "Vedaśrī", F-395, Sudarshan Park
ESI Hospital Metro Station, New Delhi - 110015
Phones: (011) 4172 2626, 98713 94815
e-mail: indology@dkprintworld.com
Website: www.dkprintworld.com

Preface

INDIAN culture and traditions derived from a comprehensive philosophy of the Vedas and other ancient texts contain many astounding feats in science and technology. Leaving aside many mythological stories there are ample archaeological, linguistic, textual, and practical evidences of a robust scientific advancement in ancient India. Elaborate brick buildings, with drainage and hygienic facilities existed over 5,000 years ago. There is evidence for large ships built in India for trade with the Middle Eastern countries. Medical science was quite advanced at least 2,500 years ago, as outlined in *Caraka Saṁhitā*, and Yoga and psychological sciences were taught as well-established subjects about 3,000 years ago. Given that the number and decimal systems were developed in India, one just has to only wonder what may have been the need at that time for these numerals. Advances in astronomy and applied mathematics have been documented within the past 2,000 years. Ayurveda and metal chemistry appear to have utilised the common knowledge of herbal chemistry for developing efficacious drugs as much as sound metal structures.

The questions frequently asked are what happened to this knowledge? If these scientific advances existed then how did they disappear? How to trace them back? The answers to such questions are complex, yet obtainable. There are historical, political, social, cultural, and economic reasons for non-availability of the records of the ancient knowledge systems. However, given the vast amount of literature available in India in Sanskrit and other languages, it is not insurmountable to trace the old scientific and technological knowledge, and weave through that ancient knowledge coherently to not only record the existence of such knowledge but also to build up theories and practices based on such knowledge under the tutelage of modern science and technology. The challenge is to look for the original manuscripts, translate them into English, and translate them scientifically, so that the material becomes comprehensible within the modern scientific paradigm.

This challenge was thrown to the Center for Indic Studies at UMass, Dartmouth, by a prominent Indian American, Mukesh Chatter, whose BVM Foundation provided initial seed funds to explore this project. Overall goals of this initiative have been (a) rediscovery of science and technology of ancient India based on original Sanskrit texts, (b) exploration of several topics of research, establishing historical validity of the texts; translating original texts from various sources, and explaining this scientific and technological knowledge in modern scientific terms for testable hypothesis, (c) creating a group of scholars, who are collectively knowledgeable in Sanskrit with a strong background to interact with modern scientists for proper translation of the texts and modern understanding of the subject matter, (d) identifying the places of high learning in India and elsewhere that may be holding academic resources and interest to collaborate on discovering the practice of science and technology in ancient India, and (e) incorporation of knowledge developed through this initiative into a sustainable living curriculum through books, journals, articles, courses, seminars, and discussions that meet the practical needs of the students and the society.

Resources were, in part, used to recruit a Mukesh and Priti Chatter Distinguished Professor of History of Science. Girish Nath Jha, a computational linguist, from the Special Centre for Sanskrit Studies at Jawaharlal Nehru University (JNU) was appointed as the first Mukesh and Priti Distinguished Professor for the project.

The centre under the leadership of Dr. Jha organised a one-day seminar on "Science and Technology in Ancient Indian Texts (SATIAIT)" on 18 June 2009 at Dev Sanskrit Vishwavidyalaya, Haridwar, inviting several experts in the field of scientific exploration of Indian traditions and practices.

That group recommended an open national seminar to seek inputs from various fields — Sanskrit, philosophy, history, computer, physics, chemistry, engineering, etc. The seminar was organised on 9-10 January 2010 at JNU, which attracted several prominent scientists, engineers, mathematicians, Sanskrit experts, and philosophers.

Although the articles being published in this volume are much less in number than those presented, the collection however, adequately represents various areas of fundamental science, medical science, philosophical approaches to science, mathematics, engineering, and technology. The challenge remains at two levels: (i) How to develop the current understanding of the ancient knowledge into modern science and technology? (ii) How to devise a fundamental level of scientific principles of the ancient knowledge into modern scientific paradigm? A lot more research work, discussions, and conferences would be needed to advance these ideas. Some aspects, such as Ayurvedic medicine and yoga sciences, are more acceptable and practical than others, such as the metallurgical advances described in *Vaimānika Śāstra*. However, the critical point in this effort is to begin collecting various ideas to examine their historicity, systematic description, and potential for theorisation so that the scientific and technological elements can be understood and applied through modern frameworks of research and development.

There are certain elements of observations made by several authors, which are encouraging for an eventual formulation of historical and practical ideas based on the ancient knowledge. For example, there is strong evidence of metallurgical advances using herbal chemistry, both in the Ayurvedic preparations of *bhasma*s and also in the development of materials and alloys used in the making of machines and structures. A real evidence of this type of technology, still available in India, is the iron pillar in Delhi that was constructed in CE 460. Once established with historicity and the underlying scientific knowledge, these examples of technological advancements are no different than what we see in modern times of the connection between modern medicine and material science, including biomaterial science and technology.

While the larger challenge of creating more comprehensive collection of testable science and technology in ancient India remains to be addressed, the proceedings of the first national seminar on SATIAIT contained in this compendium presents a robust beginning which should become a foundation for the future progress. Several initiatives have already been taken by the Center for Indic Studies, including translation of Sanskrit texts, workshop on developing modern science from Indian texts, and experimental work on science, technology, and medical applications.

We are, indeed, grateful to many individuals, including P.N. Srivastava, former vice-chancellor of JNU for his inaugural address at the seminar, Rajendra Prasad, then Rector, JNU, who welcomed the delegates to the conference. Both Srivastava and Prasad are well-accomplished life science scientists. We are also thankful to Mrs. Swarnlata, Head, Technology Development for Indian Languages programme, Department of Information Technology, Government of India, whose inaugural speech was encouraging for the scholars in linguistics as the government has drawn out plans to make much of the information available in the Indian languages digitally.

In addition to our heart-felt gratitude to Mukesh and Priti Chatter, we would like to thank colleagues at the JNU, especially those from the Special Centre of Sanskrit Studies and Centre for Philosophy, whose association was critical for this programme, and who continue to be involved in this initiative. Girish Jha, Umesh Kumar Singh, and Diwakar Mishra deserve a special mention not only for their leadership in organising the seminar, but also for their great contribution in the editing process of this volume. Finally, we acknowledge the Center for Indic Studies at UMass, Dartmouth, for its unflinching support to this project.

Dartmouth, MA **Bal Ram Singh**

Introduction

It is with profound pleasure that we are presenting this compendium of highly informative, research based, insightful presentations, on India's rich scientific heritage to our esteemed readers, anywhere and everywhere. These presentations were made by the respective author-scholars at the two-day seminar titled 'Science and Technology in Ancient Indian Texts (SATIAIT), organised at the Special Centre for Sanskrit Studies', Jawarharlal Nehru University (JNU), New Delhi, on 9-10 Janurary, 2011.

A brain-child of Prof. Bal Ram Singh, Director, Center of Indic Studies, University of Masachusetts, Dartmouth, US, it was a laudable effort for bringing together scholars from various sub-disciplines under one roof to delve deep and take stock of what we have inherited and how to preserve and carry it forward to the new horizons of glory, now on.

An advisory committee of eminent scholars was constituted to guide us in the organisation of this seminar and plan collaborative initiatives after the seminar. The following scholars kindly accepted to be on our advisory committee:

- **(Mrs) Swarn Lata**, HoD, Technology Development for Indian Languages (TDIL), Ministry of Communications and Information Technology (MCIT)
- **Radhavallabh Tripathi**, Professor and Vice-chancellor, Rashtriya Sanskrit Sansthan, New Delhi
- **K. Ramakrishna Rao**, Professor and Chairman, Indian Council of Philosophical Research (ICPR), New Delhi

- **Rajendra Prasad**, ex-Rector, JNU, and Professor, School of Life Sciences, JNU
- **V.K Jain**, Registrar, JNU, and Professor, School of Environmental Sciences, JNU
- **Vachaspati Upadhyaya**, Professor and Vice-chancellor, Sri Lal Bahadur Shastri Rashtriya Sanskrit Vidyapeeth, New Delhi
- **M.M. Agrawal**, Professor and Head, Dept. of Sanskrit, Dean, Faculty of Arts, University of Delhi
- **Bal Ram Singh**, Director, Center for Indic Studies and Botulinum Research Center, University of Massachusetts, Dartmouth, USA
- **Michel Danino**, Swarnambika Layout, Ramnagar, Coimbatore, TN
- **Shukla Mukherjee**, Associate Professor and Project Officer, Rashtriya Sanskrit Sansthan, New Delhi
- **Kapil Kapoor**, Professor and ex-Rector, JNU
- **Sankar Basu**, Professor and Dean, School of Languages, Literature and Culture Studies, JNU, Chairperson, Special Center for Sanskrit Studies, JNU
- **Shashiprabha Kumar**, Professor, Special Center for Sanskrit Studies, JNU
- **Girish Nath Jha**, Associate Professor, Special Center for Sanskrit Studies, JNU, (Mukesh and Priti Chatterjee Distinguished Professor of History of Science, Center for Indic Studies, University of Massachusetts, Dartmouth, for SATIAIT)

The seminar hosted experts, both from India and abroad, working on natural sciences in Sanskrit, under the following broad categories:

- Biological and Environmental Sciences

- Chemical Sciences and Metallurgy
- Physical Sciences and Astronomy
- Medical Sciences, Āyurveda, Yoga for Healing
- Mathematical and Linguistic Sciences
- Engineering including Vāstu, Vimānaśāstra, Ship-building, War, Armament
- Agricultural Sciences and Animal Husbandry
- Other Novel Sciences (not covered under 1-7 above)

There were a total of 11 sessions, 29 invited speakers, 6 inaugural session speakers, 4 plenary session speakers, 8 session-chairs, 11 co-ordinators and 4 panelists. Some speakers/participants could not reach the venue due to the last minute logistical problems.

Papers were presented in the following areas:

- Chemical sciences, metallurgy (Vimānaśāstra)
- History/philosophy of science/conceptual
- Language Technology/formal treatment of language
- Language (phonetics/acoustics)
- Maths
- Mechanical Engineering
- Physics
- Astronomy
- Āyurveda
- Yoga and cognitive sciences
- Law
- Knowledge management (by the panelists)

Papers in the following areas were missing:

- Vāstu
- Ship-building, war, armament, surgery
- Formal linguistics
- Agriculture

The present volume consists of 17 papers from the seminar. There was a time constraint which had to be adhered to by the authors. As a result we could get only these papers finished by them in time.

In the series of papers the first paper belongs to Prof. Bal Ram Singh titled "Scientifying the Science: Art of Making Everything a Science". Here he deliberates on the development process of science.

Main topics discussed in this paper are about the origin of universe, evolution of man, tools of understanding, consciousness as a universal concept, science as a predominant modern concept, assumptions and hypothesis, typical scientific methodology, theory of ideal gases in modern thermodynamics, language of science and relation between philosophy and physics.

The second paper belongs to Prof. R.P. Singh, titled "Spirituality, Logic and Science in India: A Philosophical Perspective". This paper is divided into three parts. In the first part Dr. Singh has discussed *parā* and *aparā-vidyā* along with *niḥśreyas* and *abhyudaya* and he says that *parā-vidyā* in ancient india was the knowledge about spirituality and *aparā-vidyā* was the knowledge related to science.

In the second part of the paper Dr. Singh talks about the notion of absolute spirit, and in the third part he has discussed Indian perspective about the spirit and some *mahāvākyas* of Indian philosophy.

The third paper is "Vaiśeṣika Insights into Science"

authored by Prof. Shashiprabha Kumar. The paper of Prof. Kumar intends to highlight the select theories of Vaiśeṣika which may provide significant inputs for different domains of science. But the paper does not, in any way, claim to be exhaustive; it is selective in its scope.

In this paper she has discussed about the basic theme of Vaiśeṣika philosophy and its relevance to the theory of matter, theory of chemical action, theory of causation, theory of metals, theory of dream and theory of motion in light of modern science.

The fourth paper is "Fourfold Description of Sound: A Vedic View" written by Prof. M.G. Prasad. In this paper Prof. Prasad says that the acoustics is a field of science and arts dealing with generation, transmission and reception of sound and acoustics have received a very high importance in the Vedas and Vedic literature.

Dr. Prasad has thoroughly explained the Vedic description of sound or speech as *parā, paśyantī, madhyamā* and *vaikharī*. He underlines the fact that the oral tradition of the Vedas does not mean that there was no writing technique in existence. Expressing that anything or any knowledge cannot be transmitted better than in a vocal and physical presence, even in the modern age of computers, that is why the Vedic *ṛṣi*s emphasised on the vocal tradition of chanting the *mantra*s.

"Relevance of Vedic Physics in the Contemporary Society" is the fifth paper in the series of seminar proceedings authored by Dr. Sadashiv Biswal. He demonstrates in this paper that besides, religion. philosophy and mythology, the Vedic literature contains various other branches of knowledge, such as astronomy, astrology, cosmogony, cosmology, space-time continuum, mathematics, chemistry, physics, meteorology, seismology, botany, zoology, geology, medicine, agriculture, environmental science and ecology and above all elements of science and technology. Issues discussed in his paper are Vedic theory of creation, Vedic theory of gravitation, Vedic theory of relativity, Einstein's mass-energy relation, geomagnetic field and geothermal field, unification of forces, space-time continuum, Vedic way of calculating the speed of light, astronomy, the atmosphere in the Vedas.

The sixth paper belongs to Dr Soma Basu titled "Concept of Agni in the *R̥gveda* in the Light of the Physical Sciences". In the first part of this paper Dr Basu has attempted to demonstrate the fact that it is through a proper understanding and realisation of history that the inherent branches of science reveal themselves. In the second part, she has demonstrated how scientific was the vision of the people of the ancient past who not only looked into life and the world with the eye of a poet-artist but also brought to light modern and advanced ideas through simple observation expressed in simple language that wondered the world in those days of the hoary past.

The seventh paper, written by Dr Premendra Priyadarshi is "Mechanics in Hindu Literature". In this paper Dr Priyadarshi, has focused on two main topics; first is Vaiśeṣika *darśana* and Praśastapāda, who was the well-known commentator of Nyāya school of philosophy. In this paper, some *sūtra*s of Nyāya-Sūtra have been examined to find out whether they meet modern standards of physics. He has examined a very limited section of the text, from which, it becomes obvious that Kaṇāda had knowledge of velocity, momentum, gravitation, acceleration, energy, etc. He says, the text anticipated Newton's Laws of Motion. It also discusses active and passive work, potential energy, collision, impulse and inertia. There is explicit mention of magnet and magnetic compass in Kaṇāda's text, which other authors have so far thought came to India only in the sixteenth century.

Paper written by Prof. Bhudev Sharma is eighth in this series and titles "Indian Mathematics and Its Applications in Vedas and Other Sanskrit Texts". Prof. Sharma in this paper says that languages and mathematics make human expression possible and exact. While languages vary, mathematics in its development is universally the same. But in India, Vedic literature played a significant role in the development of mathematics. He maintains that the birthplace of mathematics, specially that of numbers, geometry and algebra is India. Origin of numbers, geometry and sequences and series is enshrined in the Vedas. Mathematics developed in the Vedic tradition in India and from here went to other parts of the world. He emphasises that mathematics as a philosophical discipline and as a tool for understanding anything was well understood by ancient Indians.

Prof. Sharma has well discussed the history of mathematics in the world and sought its root in the Vedic literature.

Ninth paper belongs to Prof. Nabanarayan Bandyopadhyay and is titled, "An Unpublished Commentary of the Kātyāyana Śulbasūtra. It depicts about an unpublished commentary by Gaṅgādhara Pāṭhaka (seventeenth century). In this paper he speaks about the importance of this manuscript and also focuses on the fact that this commentary contains more than 50 geometrical figures with interpretation.

He says that in this commentary, many charts, diagrams, calculations, etc. relating to mathematics, especially geometry and mensuration, which are conspicuously absent in other commentaries, have been explained in a systematic and scientific way long before the advent of the European style of interpretation in this regard. He emphasises that such texts are our proud possession as relies of ancient Indian science. He advocates that we should collect, preserve, edit and publish these fruits of rich Indian intellectual tradition.

The tenth paper in the series is authored by Dr Dillip Kar; its title is "Rasaśālā, the Ancient Indian Pharmacy: A Study on its Construction". In this article he has dealt in with ancient Indian pharmacy with reference to the book *Rasāyanasāra*, written by Pt. Śyāmasundara Vaiśya in 1915. He has presented a good research work on several furnaces and other tools used for melting and making *bhasma* of all metals. These tools are named by *sattvapātanakoṣṭhī, tiryakpātanakoṣṭhī, pātālakoṣṭhī, bhūmikoṣṭhī, gārakoṣṭhī* and *aṅgārakoṣṭhī*, etc.

At the end of the paper, Dr Kar informs that to restore the ancient Indian wisdom, which is presently available in the form of manuscripts, the Government of India has established the National Mission for Manuscripts, in February 2003, and is trying to document all such rare texts like *Rasāyanasāra*, etc. He further says that it is observed that till date about 38,000 Āyurvedic manuscripts, including Vṛkṣāyurveda (plant science) and Mṛgāyurveda (veterinary science) are already documented, out of which 1,200 (1,25,000 pages) Siddha manuscripts are already digitised. When this documentation work is completed, soft copies of all texts will be made available to the scholarly world.

The eleventh paper, written by Dr C.S.R. Prabhu is titled "High Technology in Ancient Sanskrit Manuscripts". In this paper Dr Prabhu has presented his significant research works related to Vaimānikaśāstra of Bhāradvāja.

Dr Prabhu says that Bhāradvāja's *Vimānaśāstra* deals with advanced metallurgy, material science, machine design, mechanical engineering and rocketry. The text describes detailed procedures in the preparation of several hundreds of materials such as alloys and glasses which are yet unknown to modern science. Dr Prabhu claims that most of these materials can be reproduced in the laboratory even now and he too has developed some such alloy materials. The materials developed or deciphered according to Vaimānikaśāstra by Dr Prabhu are *tamogarbha loha, pañca-loha, arāra tāmra, capala grāhaka, capala grāhaka* (glass), *ravi śakti apakarṣaṇa darpaṇa* (glass), *uṣṇa śakti apakarṣaṇa darpaṇa* (glass), *badhira loha, vidyut darpaṇa, raja loha, rudantī maṇi, rutika maṇi, abhra mṛd darpaṇa, sunda mṛt kāca,* etc. Thus Dr Prabhu has done excellent job in revealing the progress of ancient Indian science and technology.

The paper (twelfth) presented by Dr Shashi Tiwari is "Environmental Science: Vedic Perspective". In this paper she has presented an overview of, Vedic thinking about nature and environment. According to her, Vedic vision on environment is well defined in verses of the *Atharvaveda* where three coverings of our surroundings are referred as *chandāṁsi*. Vedic *r̥ṣis* studied nature's drama very minutely and the *śānti mantra* of Veda is an example of this approach.

The paper titled "Contribution of Ancient India in Production Technology and Mechanical Engineering" (chapter thirteen) by Chandrabhan Prajapati presents a general introductory survey of the vaulable resources for production technology and mechanical engineering contained in the ancient texts of India.

The paper of Dr M.K. Sridhar (fourteenth in sequence) is entitled "Personality Traits of People Born in Different Saṁvatsaras and Their Comparison with Modern Psychological Personality Traits: A Study". In this paper Dr Sridhar has presented a good survey to validate his claims.

The fifteenth paper is authored by Vaidyaratnam Dr R. Raghavan and titled "The Applicative Wonders of Āyurveda: The Mother of All Life Sciences". In this paper Dr Raghavan presents the infinite possibilities of the Āyurveda as the mother of all life sciences. He realfirms the applicative aspects of Āyurveda by illustrating a number of seemingly incurable diseases which Āyurveda has successfully handled. Āyurveda, with its patient centric approach and Vedic roots as a firm foundation, has remedy for any malady. Dr Raghavan concludes that Āyurveda should provide the suffering mankind the safest remedy for leading a productive life and achieving the ultimate goal of life — *mokṣa* (liberation).

Another paper (sixteenth) is also related to Āyurveda. It is authored by Prof. Yamini Bhushan Tripathi and titled "Āyurvedic Medicines: Ancient Roots and Scientific Validation". In this paper he has quoted the basic constituents of human body — physical (*vāta, pitta, kapha*) and abstract (*sattva, rajas, tamas*). According to him the advantage of the system is that the medicines prescribed are also described in these terms only and not in terms of diseases which support the previously mentioned subjective cure methods. He has also briefly described the *rasāyanas*. He has examined the claims of some medicine manufacturers that their medicines were nano materials. He suggests that antibiotics are taken for fast actions but that this is not a concept found in Āyurveda.

Seventeenth paper of this proceeding is written jointly by (Mrs) Swarn Lata and Manoj Jain titled "Unicode Standard for Vedic Sanskrit: Efforts and Status". In this paper authors have discussed about their efforts to preserve the Vedic literature with the help of modern technology. Because of their efforts, various characters have been accepted for encoding for the representation of Vedic characters and symbols, as recommended by the Department of Information Technology, Ministry of Communications and IT, Government of India, for inclusion in the future version of ISO10646 and Unicode Standard.

The last paper by Pt. Ramsamooj "Sanskrit, Whole Brain Learning: And its Importance in the Super Accelerated Learning Theory (SALT)" highlights a very interesting and innovative idea on information management that human brain may be engaged in. It also focuses on the role of natural languages — particularly the Sanskrit language — in such interfacing. Pt Ramsamooj is considered the father of the so-called Super Accelerated Learning Theory (SALT) which appears to be very promising project at this stage.

Contents

PART III: MATHEMATICAL SCIENCES

PART IV: CHEMICAL SCIENCES, ENVIRONMENT AND METALLURGY

PART V: YOGA AND ĀYURVEDA

Part - I
Conceptual

1

Scientifying the Science
The Art of Making Everything a Science

Bal Ram Singh

Abstract

Modern science developed over the past five centuries has made major advances in the society. However, it has been dominated by materialistic ego-driven philosophy in which uniformity of observations is key to collect and interpret data. Diversity of the nature is generally categorised to describe uniform observations, leaving individual differences to be set aside in the interest of advancing empirical knowledge. Generally, there is no cause or mechanism of diversity available, which leaves a lacunae in the rational understanding of the origin and journey of the universe and life. An attempt has been made in this article to introduce Dharma paradigm to describe observations detached from the supremacy of the human existence. An example of the evolution of organism has been taken to elaborate the role of *karma* for the creation of diversity, which has been contrasted with the complexity-based evolution generally believed to explain the origin of species proposed by Charles Darwin.

Science and Dharma

ENOUGH scientific research has come out that the Yoga exercises and concepts are now being institutionalised in education, medical practice, and even in correctional facilities. For understanding the *dharma* paradigm, I believe the scientific language is the most appropriate means of communication for the world at large. Whereas science may not have solved all the problems in the world, the scientific approach has become the universally accepted way of learning and understanding. Moreover, the tradition of *dharma* is very amenable to scientific inquiry and examination, and its rational and logical features openly embrace scientific queries, without being offended. At the same time, one does not have to verify every idea or practice in *dharma* or other traditions with science to consider them valid. I only plead that the language of science and scientific approach be used in the debates and discussions about *dharma* traditions.

In fact, it is possible to use many of the concepts of *dharma* traditions to either rationalise or even reveal new concepts on some of the fundamental questions being addressed in science. These include origin of matter and universe, origin of life, origin of species and evolution, state of consciousness and mind, etc.

Dhārmic Paradigm for the Theory of Evolution

Here I take an example of the origin of species (or diversity) and evolution to explore how *dharma* paradigm can be used to rationalise and even reveal new concepts in this very contentious discipline of scientific inquiry.

One of the major features of the Darwinian theory of evolution involves the presence of diversity within and between genera of organisms from bacteria to mammals. While the basic mechanism of evolution of different species (or diversity) is explained by assuming random genetic mutations, some of which allow better adaptation to the nature, and thus natural selection resulting into continuous stream of new species, it is not known how the random mutation is initiated and what is the role of the nature in creating those mutations and selecting the species. Different *karma* aspects emphasised in various dhārmic traditions can be contrasted with modern scientific and philosophical concepts to elaborate them to the modern world, and explain some of these unresolved questions of interaction between the nature and organisms.

Karma as the Basis of Diversity

The concept of *karma* being responsible for the variation in human behaviour and their state in the society explained in Buddhist philosophy can be contrasted with the concept of evolution to clear concepts, and address missing points of even scientific theories. In Buddhist philosophy, it is the difference in *karma* which determines the birth in a given state of beings, high and low, base and exalted, happy and miserable, beings as beautiful and ugly, high-born or low-born, well-built or deformed, or the worldly conditions of beings, such as gain and loss, and disgrace, blame and praise, happiness and misery.

Clearly these concepts refer not just the physical existence, survival, and evolution. These refer to physical, mental, and intellectual aspects of a being. While there is a general concept that mental and intellectual features only belong to humans or some other higher animals, modern scientific evidence and ancient dhārmic traditions place intelligence in lowest of the low organisms (and dhārmic traditions even in non-living objects) as well, albeit not to the same degree as humans.

Inadequacy of the Current Theory of Evolution

The biological evolution, however, largely refers to the physical features of organisms, from complexities of organs and tissues to the variability in cells and size of nuclear materials. This approach counts on fossilised records on one hand and the brain size and genetic mapping on the other. While the scientific approach is considered largely objective, it has been

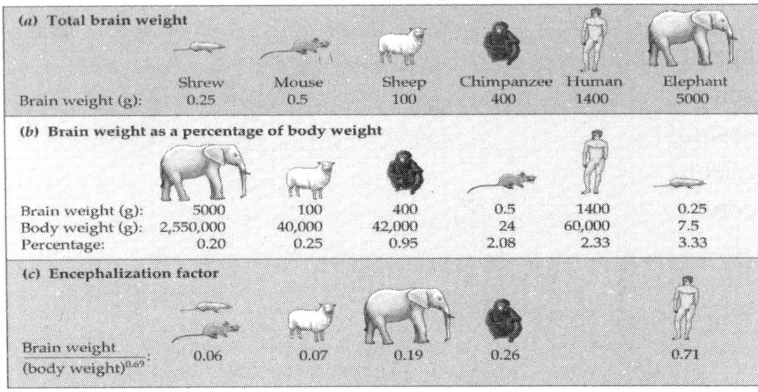

fig. 1: Different ways to calculate brain size to show evolution in different animals [Taken from *Biological Psychology: An Introduction to Cognitive Neuroscience*, 4th edn., 2004, Sinauer Associates Inc.]

marred with a certain degree of manipulation as it relates to the discussion of evolution. Whereas Darwin published his book *Origin of Species* to explain the diversity of biological organisms, and potential evolution of one from the other through the process of adaptation and natural selection, evolution itself has been taken as a trait for superiority. As shown in ***fig.* 1**, there have been attempts by "scientists" to find a way to place human beings on top of the evolutionary ladder by deriving an encephalisation factor that just fits the requirement of human beings having the largest encephalisation factor.

Similarly, an unwritten hypothesis and expectation in the human genome project was that humans will have the largest number of genes and perhaps the largest genome — both of which have turned out to be not true. Humans have about 30,000 genes whereas plants and mice have over 60,000 genes. In fact, onions, amphibians, and whisk fern, have six times, 30 times, and 3,000 times, respectively, larger DNA in their genomes than humans.

The larger issue is whether in being so biased in our approach at the fundamental level we are compromising the science. Taking the example of evolution and its basic assumptions one can see how a perspective can change the scientific explanation of an observation. The following are the core assumptions of evolution:

1. The diversity of species we see in the world basically evolved from unicellular organisms through the process of natural selection.
2. Natural selection occurs when a given species adapts to the change in nature through its variation in the biological system.
3. The basis of the variation in the biological system is random mutation in the genetic material of the organism.
4. For an organism to accumulate large number of random mutations it must undergo many generations over a long period of time.

Humans being in existence for about only 1 million years, and with their reproduction cycle of about 20 years would have gone through only about 500 iterations. On the other hand, anaerobic bacteria with a record of existence for about 2 billion years would have gone through 2.16 trillion iterations considering even 8 hours for their life cycle. The anaerobic bacteria thus have 4.32 billion times advantage over humans to go through the adaptation and natural selection process compared to humans. It is thus important to consider the bacteria being superior evolutionarily. This is especially relevant since the bacteria have survived in the nature throughout the time other organisms and species have appeared. Similar arguments can be made for plants and animals.

Although there are no written rules or expectations for humans to be most evolved in terms of superiority, it has been generally accepted that human species is the most evolved and most superior among the current species of organisms. This has sometimes led to bending or accommodating the rules to suit the human superiority.

As mentioned earlier, this idea of superiority led to the thought when the human genome project was undertaken that the human genome will be the largest, containing the largest number of genes to justify their status of most complex set of organs, physiology, cognition, etc. However, as it turned out not to be the case, scientists are still struggling with this question, and are looking for ways to justify the concept of human superiority.

In the case of bacteria, with only about 5 million base pairs of DNA and about 5,000 genes, about 80 per cent of the genetic material is utilised for genes. This compares with 1.4 per cent utilisation of the genetic material (approximately 3 billion bp) for genes (30,000). Thus on the basis of efficiency in using its genetic material, bacteria seem to be more evolved.

Bacteria, plants, and animals other than humans can readily live in the nature, thus demonstrating their adaptation and natural selection trait, whereas humans generally live an

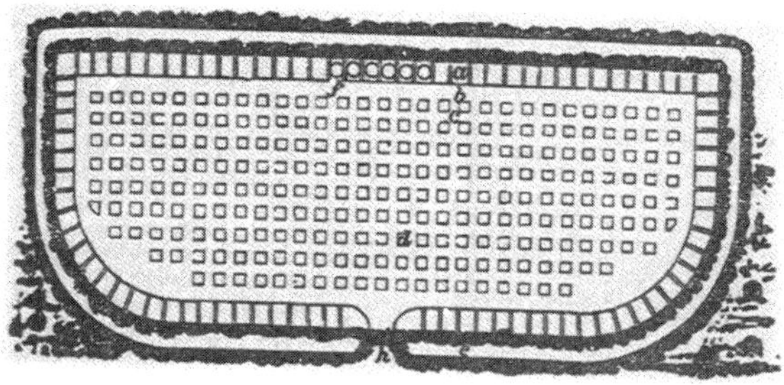

fig. 2: Schematics of 242 experimental plots — 4×4 ft; 85 beds Clover, Trefoil, Lucern, Saintfoin, etc. that were used to show that different species of grass grown together produce larger yields than when grown separately. [Schematics taken from Hector and Hooper, 2002] artificial life, which not only shows their inability to cope with the natural living, but it has also been suggested that their further evolution is in jeopardy, as there is no adequate interaction with the nature.

Darwin on Diversity

In his famous book, *Origin of Species,* Charles Darwin cited a study (Sinclair, 1826) which demonstrated that when several types of grasses (Clover, Trefoil, Lucern, Saintfoin, etc.) were planted together there was a higher yield for each compared to when each grass was planted separately. In *Origin of Species,* Charles Darwin writes,

> It has been experimentally proved that if a plot of ground be sown with one species of grass, and a similar plot be sown with several distinct genera of grasses, a greater number of plants and a greater weight of dry herbage can thus be raised.

By this he demonstrated the principle of divergence in the natural selection and evolution process. This is the key force driving the diversity even from evolutionary point of view.

What is missing in this observation is the origin of the force that drives this diversity. Both the yield and entropy (thermodynamic measure of randomness) based arguments are pertaining to the consequences, not the cause. *Karma* and the origin of *karma* (the author of the *Rājavārttika* in Jaina tradition points out that which is performed out of the result of the vibrations of the self is called *karma,* which is similarly described in Nāsadīya-Sūkta as Big Bang Singularity, the state before which everything was pure existence and pure consciousness, *anid avataṁ,* without vibration, and thus the world begins with vibration) point towards the origin of the force that drives the diversity.

Evolutionary Advancement of Bacteria and Plants

Plants and other organisms, including animals, by virtue of

being on Earth for longer time period than about 1 million years humans have inhabited this planet, one would imagine that those organisms such as bacteria which have existed for over 1 billion years are likely to have undergone higher level of evolution through the iteration of adaptation and natural selection, particularly because their reproductive cycle runs at a shorter timescale. Such a realisation is likely to be more in tune with dhārmic traditions which value and respect earlier creations to the extent that some of them even worship them. This may reflect the longer memory and observations by continual civilisations based on existence of *dharma* traditions, and their integral approach to the universe and understanding of the existence of various species and organisms more a reflection of *karma* than a particular hierarchy. There are many stories in epics and Purāṇas which suggest at times animals and plants having higher level of consciousness than humans.

Integration of Metaphysical Concepts in Science

While biological evolution does not necessarily include metaphysical concepts like mind, intellect, etc., dhārmic traditions have considered evolution as a process at different levels of our own existence evolve into the other. The idea is based on the *pañcakośa* system whereby humans have their existence from *annamaya* (physical), *prāṇamaya* (energy or air), *manomaya* (mind), and *jñānamaya* (intellect) *kośa*s (sheaths), and can evolve further into *ānandamaya* (bliss or angelic) *kośa*.

The *pañcakośa* system allows integration of humans to rest of the living and non-living matters of the universe on one hand and the state of *ānanda* on the other, again linking many concepts of practice of universalism and *ahiṁsā* for human, and something that can be attained through *karma*.

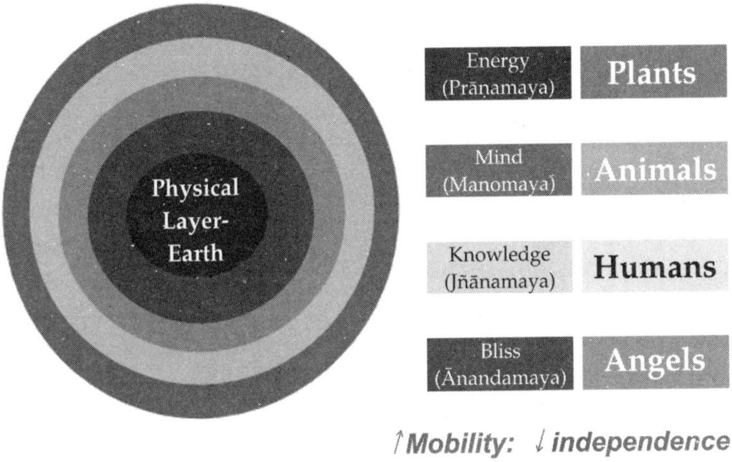

fig. 3: Pictorial representation of *pañcakośa* — *annamaya prāṇamaya, manomaya, jñānamaya* and *ānandamaya*, with primary examples where such *kośa*s are dominating features — mineral (earth), plants, animals, humans, and angels, respectively. As the evolution of matter occurs from *annamaya* to *ānandamaya*, the mobility is increased but independence is restricted. For example, earth or mountains are quite immobile but does not need other forms of matter for existence whereas angels are quite mobile but cannot exist without the rest. [Modified from Sharmon and Baginsky, 2003]

The acceptance of the physical nature of human being, in addition to the *prāṇamaya kośa*, *manomaya kośa*, and *jñānamaya kośa*, allows us to look into scientific observations with sympathy, and provide further ideas to advance or evolve the physical nature to higher levels of human existence.

Furthermore, the concept of physical and subtle bodies — ethereal, astral, mental, and spiritual, along with the concept of *cakra*s and *nāḍī*s — allows explanation of the operational aspects of reincarnation. The *prāṇa* energy which flows freely not only in the *nāḍī*s of physical body but also carries information amongst different bodies through energy centers or *cakra*s. According to the *cakra* hypothesis, at the death of a human being only physical and ethereal bodies die, leaving all the information encoded in other subtle bodies. Once the soul is reborn in a new physical body the remaining subtle bodies reconnect and actions of past lives become incorporated with the new physical body and life. The *cakra* system thus allows experiential observation through Yoga practice making it a real world observation. This system thus provides a testable concept for scientific analysis.

Fig. 3 is a pictorial representation of the existence of five types of bodies of human beings. While death occurs to physical and ethereal bodies, the remaining subtle bodies remain eternally and carry information coded in them about the action and its derived consequences from life to life.

Finally, some of the scientific ideas, such as Newton's third law can be challenged for their exclusive description of only physical nature, and can perhaps be modified to include subtle nature of human existence which can provide an integrated and more advanced form of scientific concepts. At the same time, concepts of force and their cause in *karma,* as described in Jaina philosophy above can be used to create clearer and observable ideas of *puruṣārtha.*

Puruṣārtha is something that is demonstrated in action or by *karma* rather than talking about it. That is a major element of dhārmic traditions whereby an *ācārya* demonstrates the principles by conducting oneself accordingly. This makes it easier to communicate, much like demonstration of science.

Conclusion

Combining the essential values of dhārmic traditions with several practices such as Yoga, and explaining in terms of science and scientific approach is likely to advance the ideas like *karma* to the fore of discussion and debate to the benefit of both science and society. Similarly, other issues such as origin of matter, mind and intellect, and origin and cure of diseases, etc. can be addressed by scientifying philosophical concepts originated in India over a long period of time.

References

Hector, A., and R. Hooper, "Darwin and the First Ecological Experiment", *Science* 295, 639-40, 2002.

Sharmon, S., and B. J. Baginsky, *The Chakra Handbook*, New Delhi: Motilal Banarsidass, 2003.

Sinclair, G., *London's Gardener Magazine*, 1, 14, 1826.

2

Spirituality, Logic and Science in India
A Philosophical Perspective

R.P. Singh

Spirituality, Logic and Science are regarded by many thinkers as two separate realms. Prominent among them is Kant who has said, "I must limit knowledge in order to leave room for faith". He created two realms — a realm of scientific knowledge and another realm of faith, morality and religion. What we normally call *science* is that which we can observe, experiment, analyse, measure and prove. From that perspective, *spirit* or spirituality may be or may not at all be the last thing to be found in any microscope. The science of *Spirituality* may be not so easily measured from the parameters used in normal *science*. I propose the role of Logic between Spirit and Science. The insight into this proposal comes basically from the Upaniṣads in which the distinction has been made between *parā-vidyā* and *aparā-vidyā* at the level of knowledge and between *niḥśreyas* (attainment/fulfilment) and *abhyudaya* (achievement) at the ethical level. The same *dṛṣṭi* can be found in Advaita Vedānta traditions from medieval to the modern India. However in the Western philosophy, in the post-renaissance, particularly during enlightenment, we come across a distinction between science and religion. This distinction, on the basis of principles, emerges as two separate realms in Kant. In its further development, we find that Hegel's *Science of Logic* places Logic as the middle term between Spirit and Science. This is the position of Absolute Monism which is very close to Advaita Vedānta.

LET me first see the status of spirituality and science especially with reference to modern India. India is world's largest democracy and has a long history of democratic functioning with pluralistic values and philosophic universal concepts (holistic values included). As a matter of fact, the ancient world was finite in terms of its socio-economic conditions, scientific development, historical and cultural products. The modern world was infinite in terms of ideology, scientific and technological development, logocentrism, foundationalism, essentialism, and teleology, unified world-order, rationality, conceptions of morality and justice, etc. The post-modern world has again shrunk into finite propositions in terms of anti-foundationalism, anti-essentialism and anti-teleology, fragmentation, irrationality and plurality of ethnic identities, linguistic identities, etc. In the ancient world, the Indians, the Chinese, the Greeks, lived without much contact with one another and in that sense there was pluralism. But each culture regarded its principles to be universally valid. The *ṛta* of the Vedas, the Platonic *Forms*, the *Tao* of the Chinese — all claimed to be universal. Without their knowing it, they agreed that as in the cosmos, so in the human order *yathā piṇḍe, tathā brahmāṇḍe*. Today the world is one; the Chinese, the Indians, the Europeans or the Americans mingle in academia and in the marketplace.

Modern science and technology has played the most important role in bringing the people so close to each other. There have been three stages of the development of modern science in the last century — the development of Quantum Mechanics in the 1920s; the development of molecular, cell and DNA Biology through the 1980s, the development of computers, information, automation, brain sciences and human genome, all in a piece through the 1990s. Modern technology, far from being more labour saving, has taken man out from the Planet Ocean and has hinted at the possibility of establishing interplanetary relations. We need not only to comprehend this technology philosophically but also to participate with it particularly with the rise and development of modernity. Science, technology and philosophy have always been complementing one another. The works of Albert Einstein, Arthur Edington, Werner Heisenberg, Julian Huxley, Erwin Schrödinger, Neils Bohr and others support this contention. Modern philosophy needs scientific foundation to give a deeper and more correct explanation of natural and social phenomena. Science also needs philosophical justification to relate and evaluate the technological products and scientific theories with the cultures and values on the one hand and with the aspirations and goals of the people on the other. This, however, has given rise to the problem of cultural relativism, which today in the post-modern world reigns supreme.

Morals and conceptions of good are said to be relative to cultures, even large cultural identities have now disintegrated into smaller ones — multiculturalism. Moral theories still oscillate between utilitarianism in so far as public policy, decisions of the emerging democracies are concerned, and Kantianism, in so far as individual moral life and principles are concerned. Both of these alternatives, however appealing, are now seen to be of limited application. More attractive today are Aristotelian *Ethics* expanded to include many present-day virtues, and, a communitarian ethics which goes back to a sort of Hegelian *Sittlichkeit* as the basis of one's moral conceptions. Both of these allow for pluralism and reject a monistic theory. Pluralism is the ethos of the postmodernity. On the one hand, there are the pluralities of nation-states, of world religions, of large linguistic groups, with numerous internal differences; on the other hand, there is an overwhelming sense of ONE WORLD, i.e. the world coming together through technology. The idea of a GLOBAL VILLAGE is an ideal, which is widely valued. The Vedic exhortation of *ekam sat viprāḥ bahudhā vadanti* is the need for the post-modern philosophy. The Reality admits of alternative approaches in terms of thought constructions and linguistic expressions.

Let us go into the details of those spiritual truths. Swami Vivekananda has said, "Truth does not pay homage to any society, ancient or modern. Society has to pay homage to the truths or die".[1] On the cognition of the spirit as the ultimate reality, the *Taittirīya Upaniṣad* says: That, from which our speech turns back along with mind, being unable to comprehend its fullness, is the Ultimate Reality. *Kenopaniṣad* says: That where the eye is unable to go, where neither speech nor mind is able to reach — what conception can we have of it, except that it is beyond all that is known, and beyond all that is unknown. Śaṅkara and Hegel would agree with each other that the comprehension of Being as immanent in all other categories is the most fundamental category of all things. Śaṅkara in his *Commentary on the Chāndogya Upaniṣad* (VI.2.1) says, "Existence or Being, it is urged, is pure, subtle, indefinable, all-pervading, one, taintless, indivisible, knowledge." On the concept of Pure Being, Hegel has also to say the same thing: Pure Being is complete indeterminacy, absence of all qualities, a vacuity, an empty concept. Being is the most irrepressible category for both Hegel and Śaṅkara. Hegel's Spirit is the unity, which realises itself in the differences, and it is not in which all differences are lost like that of Schelling. For Hegel, nothing could be external to Spirit; but as Spirit exists through its opposition to the "other" which it at the same time sublates and this process goes on until it reaches to the Absolute Unity. Whether this state of Absolute Monism is to be intellectually apprehended as Hegel does or expressed in self-realisation as Śaṅkara does, both the ways are of the same spiritual pilgrimage. I propose to develop this paper in accordance with following Parts:

1. Swami Vivekananda, *Collected Works* (Calcutta: Bellur Math, 1982), vol. II, p. 1.

Part 1
Parā-vidyā (Spirituality) and Aparā-vidyā (Science) Niḥśreyas (Attainment/Fulfilment) and Abhyudaya (Achievement)

The Upaniṣads draw the distinction between the *parā-vidyā* and *aparā-vidyā*, between the lower and the higher knowledge as the Greek philosophers did between *Doxa* and *episteme*, between opinion and truth. *Muṇḍakopaniṣad* tells us that there are two different kinds of knowledge, one the higher and the other the lower knowledge. Of these the lower knowledge is the knowledge of the Vedas, of grammar, of etymology, of metres, of the science of the heavens; while the higher knowledge is that which alone reaches the imperishable Being.[2] The same typical distinction between the way of knowledge and the way of realisation is brought out in a dialogue between Nārada and Sanatkumār, where Nārada, the spiritual disciple, goes to his teacher to learn the science of realisation. Asked to say what branches of knowledge he has hitherto studied, Nārada tells Sanatkumār that he has studied all the Vedas, as well as all history and mythology; he has studied the science of manes, mathematics, the science of portents, the science of time, logic, ethics, the science of gods, the science of *Brahman*, the science of demons, the science of weapons, astronomy, as well as the science of charms and fine arts. But he tells his master that grief fills him that so much knowledge is not competent to land him beyond the ocean of sorrow. He has studied only the different *mantras*; but he has not known the self. He has known erstwhile from persons revered like his spiritual teacher that he alone is able to go beyond the ocean of sorrow who can cross it by the saving bund of *ātmaṇ*. Would his spiritual teacher enable him to cross over the ocean of

---

2. *Muṇḍaka Upaniṣad*, I.4.5, tr. Vidyavachaspati V. Panoli, *Upanishads in Śaṅkara's Own Words*, vol. II (Calicut: The Mathrubhumi Printing and Publishing Co. Ltd., 1992), p. 117.

ignorance and grief?[3] This passage brings out the difference between lower and higher knowledge. Finally the extremely practical character of the Upaniṣadic seers towards the problem of self-realisation is exhibited in the *Kenopaniṣad*, where we are told that the end of life may be attained only if the self were to be realised even while the body lasts; for if self-knowledge does not come while the body lasts, one cannot even so much as imagine what ills may be in store for him after death.[4] The same idea is urged with a slightly different emphasis in the *Kaṭhopaniṣad*, where we are told that unless a man is able to realise Self while the body lasts, he must have to go from life to life through a series of incarnations.

Part 2
The Notion of Absolute Spirit

If we look at the history of the problem of the Ultimate Reality or the notion of Absolute, we find that there are three ways that we can approach to this problem — the psychological, the theological and the cosmological. We can begin with our analysis into what goes on in our mind or the *antaḥkaraṇa* and try to find the absolute in us. Or we can begin with an estimation of the outside world and try to establish an absolute causality operating in nature. Or finally, we can resort to God as the ultimate unifying force between the inner world and the outer world. The Father of Modern Philosophy, Rene Descartes (1596-1650), in search of certainty in philosophy, started with a method of doubt regarding the outside world and arrives at the certainty of the *cogito*. From the *cogito*, says Descartes, we arrive at the conception of God, who is the cause of the *cogito*, and whom we must, therefore, regard as

3. *Chāndogya Upaniṣad*, VII.1.3, tr. Vidyavachaspati V. Panoli, *Upanishads in Śaṅkara's Own Words*, vol. III, (Calicut: The Mathrubhumi Printing and Publishing Co. Ltd., 1993), pp. 693-96.

4. R.D. Ranade, *A Constructive Survey of Upanisadic Philosophy* (Bombay: Bharatiya Vidya Bhavan, 1968), p. 240.

more perfect than the *cogito.* Finally, it is from God that he arrives at the world, which he had started by negating, by regarding as the creation of a demon. Opposed to Descartes, to the God-intoxicated philosopher, Spinoza (1632-77), neither the mind nor the material reality is primary. To him, God is the be-all and the end-all of all things; it is self-conceived, self-existent and self-caused. Even Kant, to anticipate a later discussion, has subscribed to Rational Psychology, Rational Cosmology and Rational Theology. However, the way the Upaniṣads deal with the problem of Absolute is neither the Cartesian nor the Spinozistic. Śaṅkara, for existence, subordinated the world and the God to the self.

Part 3
Indian Perspective

Anti-foundationalism in the Nāsadiya-Sūkta of *R̥gveda,* 10th *maṇḍala,* R̥ṣi Dīrghatamas says

> At first was neither Being nor non-Being. There was not air nor yet sky beyond. What was its wrapping? Where? In whose protection? Was water there, unfathomable and deep? There was no death then, nor yet deathlessness; of night or day there was not any sign. Who really knows? Who can presume to tell it? Whence was it born? Whence issued this creation? Even the gods came after this emergence. Then who can tell from whence it came to be? That out of which creation has arisen, whether it held it firm or it did not, He who surveys it in the highest heaven, He surely knows — or maybe He does not know!

Issues of Contestations/Discrepancies vis-à-vis Scepticism/Anti-foundationalism

1. The basic presupposition — *mahat, prāṇa* and *ākāśa.*
2. From unmanifest to the manifest.
3. Consciousness — Ontological, Metaphysical, Epistemological, Psychological, Moral and Yoga.

4. Degrees of Reality — *pārmārthika, vyāvahārika* and *prātibhāsika.*
5. Hermeneutics — Self is within, self is without; self is before, self is behind; self is on the right; self is on the left; self is above and self is below.
6. Distinction between *parā-vidyā* and *aparā-vidyā "brahma satyam jagat mithyā jīvo brahmaiva na aparaḥ"* (*Brahman* the Absolute alone is real; this world is unreal; and the *jīva* or the individual soul is non-different from *Brahman*). The *jīva* and the means to *mokṣa.*

The Vedic *r̥ṣis* expressed their vision of the Absolute or the Ultimate Reality in the form of hymns, evoking responses at the varying levels of self-realisation, worshipful devotion or ritual sacrifices. The Upaniṣadic saints tried to impart the knowledge of the Absolute to his pupil through the method of a dialogue. An example of this is found in the *Br̥hadāraṇyaka Upaniṣad* in the dialogue between the sage Yājñavalkya and his wife Maitreyī who asks him question about the way to realise the Absolute or the Ultimate Truth. As the dialogue proceeds, Yājñavalkya convinces her that it is the *ātman* or self that should be known, and when this knowledge is achieved, everything in the Universe is known, because there lies an identity between *ātman* (self) and *Brahman* (the Ultimate Reality). The five *mahāvākyas* or great teachings (literally, great sentences), express this vision from different Upaniṣads in the following:

- *prajñānam brahma*: "Consciousness is *Brahman* (*Aitareya Upaniṣad*, III.1.3)
- *aham brahmāsmi*: "I am *Brahman*" (*Br̥hadāraṇyaka Upaniṣad*, IV.3.23)
- *aham eva idam sarvosmi*: "I indeed am this whole Universe." *Tattvamasi* (That Thou art), (*Chāndogya Upaniṣad*, VI.8.7)
- *ayam ātmā brahma*: "This *ātman* is *Brahman*" *Chāndogya*

Upaniṣad, III.10-14, *Taittirīya Upaniṣad,* I.5, *Bṛhadāraṇyaka Upaniṣad,* II.5.19)

- *sarvam khalu idam brahma*: "Everything is *Brahman*" (*Muṇḍaka Upaniṣad,* III.13.1, *Chāndogya Upaniṣad,* III.14.1)

The *mahāvākyas*, in brief, are the different paths for the realisation of the Absolute. The statement *prajñānam brahma* directs the seeker to meditate on the *cittasvarūpa* (Pure Consciousness form) of *Brahman,* leading to the merger of the individual consciousness in the Universal and the attainment of *mukti* (liberation from worldly bondage) and the state of Supreme Bliss (*paramānanda*), *aham brahmāsmi* is an endeavour to make the pupil engage in deep meditation to realise the Absolute, in the *tattvamasi* the teacher is trying to make his pupil realise that his innate Being is part of the Absolute, *ayam ātmā brahman* also accomplishes the same thing.

For Indian minds, puzzled about the dilemmas in spirituality and science, I will humbly recommend a change of perspective. Let us leave Aristotle, Kant, Hegel and even Marx, get out of the Enlightenment Frame of Mind and go to the Upaniṣads. There is no other way of detoxifying ourselves from the fumes of Enlightenment Rationality because the western way is not the only way of thinking and experiencing, let us as Indians emerge ourselves in our own rich Indian Heritage, especially before its breaking up into Buddhist, Jaina and Hindu — for example, the Sāṁkhya-Yoga heritage common to all three traditions — the great philosophical perspective that undergrads all Upaniṣadic, Vedic, Buddhist, or Jaina thought and experience. Keep your painfully acquired scientific rationality from Kant and Hegel, but do get not tyrannised by it. Stay critical, but do not reject out of hand what seems strange at first. Expose yourself without hesitation to a system of thought and experience, which has endured for millennia and awarded us with freedom, autonomy and sovereignty.

PART - II
PHYSICAL SCIENCES

3

Vaiśeṣika Insights into Science

Shashiprabha Kumar

The paper intends to highlight select theories of Vaiśeṣika which may provide significant inputs for different domains of science. But the paper does not, in any way, claim to be exhaustive; it is selective in its scope.

1. Introduction

VAIŚEṢIKA is held to be one among the six major Śāstras of Indian philosophy and is stated to be beneficial for comprehending all other branches of Indian learning.[1] In the Indian philosophical tradition, Śāstra has been stated as an equivalent term for denoting science or a scientific treatise,[2] since it not only enunciates the nature of reality but also enjoins human beings to think and act in a systematic manner, so that they can attain the goal of final beatitude.[3]

The system of Vaiśeṣika is classified as a "Nature-philosophical school"[4] since it mainly deals with the external

1. *Kaṇādaṁ pāṇinīyaṁ ca sarvaśāstropakārakam* (anonymous).
2. Monier Williams, 2005, *Sanskrit-English Dictionary*, Sharada Publishing House, Delhi, rpt., p. 1069 and Karl H. Potter, 2004, *Encyclopedia of Indian Philosophies*, vol. II, *The Tradition of Nyāya-Vaiśeṣika up to Gaṅgeśa*, Delhi: Motilal Banarsidass, rpt., p. 737.
3. *Śreyaḥsādhanābhidhāyakaṁ śāstram — Saptapadārthī*, p. 153.
4. Erich Frauwallner, 1973, *History of Indian Philosophy*, vol. II , English tr. by V.M. Bedekar, Delhi: Motilal Banarsidass, p. 3.

world. As compared to other philosophical schools in Indian tradition, the Vaiśeṣika stands out differently because it endeavours to explain the world around us in a realistic and scientific manner rather than the "world-negating attitude of some forms of Vedānta and Buddhism".[5]

Both, science and philosophy essentially deal with the same problem, i.e. understanding the world in its entirety, although their methods might be different. Science approaches the world through observation and experimentation while philosophy handles it through introspection and reflection. It is not without reason then that philosophy is said to begin from where physical science ends.[6] The same is duly connoted by the appellations *physics* and *metaphysics*. From this point of view, both science and philosophy are mutually complimentary because science aims at exploring the reality and formulating theories thereof while philosophy tries to expound and examine those theories.

In the above background, it is not surprising that since ancient times several philosophers have also been scientists and vice versa. Kaṇāda, the propounder of Vaiśeṣika philosophy, is one of such sage-scientists born in India (600 BCE approx.) who has harmonised physics and philosophy in his theories.[7] While explaining the physical nature of universe, he has not only discussed the material world but also tried to explain the mental aspect of reality and even touched the realm

5. B.K. Matilal, 1977, *Nyāya-Vaiśeṣika,* Wiesbaden: Otto Harrassowitz, p. 112.

6. James Jeans, 1948, rpt., *Physics and Philosophy,* London: Cambridge University Press, p. 17.

7. "The philosophy dealing with the composite structure of matter starting with *Viśeṣa* is called the Vaiśeṣikadarśana-Emperio-dialectics or Physical Science". Pandit, Kashinath Sharma, 1972, *Vaisheshikadarshana with Vedabhaskara Commentary,* Hoshiarpur: Hoshiarpur District Co-operative Union Printing Press, Introduction, p. xv.

of the spiritual. The school of Vaiśeṣika is so designated because it distinguishes itself from other schools of Indian philosophy in the sense that it focuses upon individual difference of each of the innumerable atoms of the universe rather than dwelling upon their inherent identity.[8] Obviously, such a rationalistic and pluralistic system suits the modern mindset prone to scientific reasoning.

The Basic Theme of Vaiśeṣika

The scientific outlook of Kaṇāda is visible right from the beginning of his Sūtras where he defines *dharma* as a source of *abhyudaya* (material progress) and *niḥśreyasa* (spiritual exaltation)[9] and declares that the true knowledge of six categories (*padārtha*s) propounded by him can be acquired through twin methods of knowing similarities (*sādharmya*s) and dissimilarities (*vaidharmya*s) which would enable one to obtain the ultimate goal of life.[10] This technique of blending the binary aspects of reality is a clue to the philosophy of Kaṇāda which sets in explaining the inclusive nature of reality such as the visible and the invisible, the gross and the subtle, matter and the spirit, substance and the attributes, generality and particularity, etc. The very idea of classifying the whole universe into six (or seven in the later tradition of Vaiśeṣika) types of categories (*padārtha*s) and claiming that nothing remains out of these categories,[11] then proclaiming that all of

---

8. "The word 'Vaiśeṣika' is derived from Viśeṣa, which means difference and the doctrine is so designated, because, according to it, diversity, and not unity, is at the root of the Universe". — Hiriyanna, 1932, *Outlines of Indian Philosophy*, London: George Allen and Unwin, p. 225.

9. *yato 'bhyudayaniḥśreyasasiddhiḥ sa dharmaḥ* — *Vaiśeṣika-Sūtra* (*VS*), 1/1/2.

10. *dharmaviśeṣaprasūtāt dravyaguṇakarmasāmānya-viśeṣasamavāyānāṁ padārthānāṁ sādharmyavaidharmyābhyāṁ tattvajñānānniḥśreyasam.* — *VS*, 1/1/4.

11. *etasyāmeva bhavātmakaṁ viśvamantarbhavati* — *Tārkikarakṣā*, pp. 162-63

these six categories are existent, knowable and nameable[12] — is thoroughly scientific in its approach. What is more important to note is that all this has been done by Kaṇāda and his followers purely with an intuitive vision and without the assistance of any laboratory or equipment.[13] It is through the well-practised austerities of the seers like him that they are endowed with intuitive insights into the nature of reality.[14] But at the same time, the Vaiśeṣika scholars give due cognisance to the day-to-day experience of ordinary human beings as well. That is why the system is also known as "common-sense realism".[15]

The Vaiśeṣika theories of atomic creation, chemical action, classification of metals as fiery substances, kinetic discussion of activity, elucidation of acoustic processes with reference to the origin and reception of *śabda*, physiological functions within the human organism as also the acceptance of life in plant-bodies, exposition of psychological processes like cognition and dream are a few examples of scientific insights available in Vaiśeṣika sources. An in-depth discussion of few of these theories will be the main focus of this paper.

---

12. *ṣaṇṇāmapi padārthānām astitvābhidheyatvajñeyatvāni* — *Praśastapādabhāṣyam* (*PPB*), p. 41.
13. "We might even be tempted to contrast some of the discoveries of modern chemists and physicists with the crude but shrewd ideas of Indian philosophers prosecuting their investigations more than 2000 years ago without the aids and appliances now at everyone's command". Monier, Williams, 1974, *Indian Wisdom or Examples of the Religious, Philosophical, and Ethical Doctrines of the Hindus*, Delhi: Indian Reprint Publishing Company, Ist Indian rpt., p. 81.
14. *ārṣaṁ siddhadarśanaṁ ca dharmebhyaḥ* — *VS*, 9/2/13.
15. Jonardon Ganeri, "Realism in Vaiśeṣika and Navyanyāya", *Materialism and Immaterialism in India and Europe* (forthcoming). ed. Partha Ghose, Project of History of Indian Science, Philosophy and Culture, vol. XII, part 5, Delhi: Centre for Studies in Civilization.

Theory of Matter

The most notable contribution of Vaiśeṣika system to the history of scientific thought has been its atomic theory.[16] In fact, it is due to this theory that the propounder of Vaiśeṣika school was given a specific appellation, i.e. Kaṇāda (atom-eater).[17] According to Kaṇāda, the physical world comprises of three types of reals:

(i) Matter in the form of four elemental substances (*bhūta dravyas*): earth (*pr̥thivī*), water (*jala*), fire (*tejas*) and air (*vāyu*). Atoms (*paramāṇus*) are held to be the ultimate, eternal, uncaused constituents of these four substances.[18]

(ii) Living bodies (*śarīras*) of four types: (a) embryonic like human beings and animals (*jarāyuja*); (b) egg-born like birds, serpents and fishes (*aṇḍaja*); (c) engendered by heat and sweat like mosquitoes and lice (*svedaja*); and (d) germinating from the earth like plants and herbs, etc. (*udbhija*).

(iii) Sense organs (*indriyas*), created out of the atoms of above four substances, namely: olfactory sense (*ghrāṇa*), sense of taste (*rasanā*), organ of sight (*cakṣuḥ*) and sense of touch (*tvak*) along with the eternal and uncreated minds (*manas*) as instruments of mental awareness.

It is worth re-iterating in this regard that four types of atoms are the ultimate material cause of universe in Vaiśeṣika view. The various visible products in the world are created

16. For details see Shashiprabha Kumar, "Atomic Theory of Vaiśeṣika" in *Journal of Oriental Research Institute*, Mysore (forthcoming), (Paper presented at the National Symposium on "The Science in India" organised by the National Centre for History of Science and University of Mysore during 21-24 April 2008).

17. *kaṇān attīti kaṇādaḥ* — *Vyomavati*, p. 20.

18. *Sadakāraṇavannityam* — *VS*, 4/1/1.

out of different conglomerations of these invisible atoms. More significantly, each of the innumerable atoms of the universe is distinct in itself, since it has an individual identity or *viśeṣa* of its own. The Vaiśeṣika acceptance of atoms as the ultimate, partless parts of all the visible products[19] is based on the logical postulate of avoiding *regressus ad infinitum*. Accordingly, each visible object is held to be constituted of parts, and these parts are again divisible into further smaller parts. But since this process cannot go on indefinitely, hence one is forced to set an ultimate limit and stop somewhere. This final limit of divisibility is stated to be *paramāṇu* in several source texts of Vaiśeṣika[20] and its existence is mainly proved through inferential process,[21] although it is said to be perceptible for the *yogin*s and verbal testimony is also adduced in this regard.[22]

The Vaiśeṣika theory of atoms is undoubtedly a starting-point in the history of physical concepts for explaining the universe in a causal framework. To quote a contemporary scientist,

> The ground layer consists of indivisible, invisible and indestructible atoms (*aṇu, paramāṇu*). It is the aggregation of these atoms that gives rise to different destructible compound

---

19. For details on the nature and features of these *paramāṇus*, see M.K. Gangopadhyaya, 1980, *Indian Atomism: History and Sources*, Calcutta: Bagchi Indological Series, p. 22.

20. (i) *tathā ghaṭadikāraṇakāraṇeṣvalpatarādibhāvaḥ kvacidviśrāntaḥ taratamaśabdavācyatvānmahāparimāṇavat. yatra viśrāntāste paramāṇva iti — Vyomavatī*, p. 224.

    (ii) *aṇuparimāṇatāratamyaṁ kvacid viśrāntaṁ parimāṇatāratamyatvād mahatparimāṇatāratamyavat, yatredaṁ viśrāntaṁ yataḥ paramāṇurnāsti sa paramāṇuḥ. — Nyāyakandalī*, p. 79.

21. *tasya kāryaṁ liṅgam — VS*, 4/1/2. For details see Uttamur T. Viraraghavacharya, 1958, *Vaiśeṣika Darśana with Rasāyana Commentary*, Madras: Vivekananda College, Mylapore, pp. 141-44.

22. For details see Shashiprabha Kumar, 1992, *Vaiśeṣika Darśana meiṅ Padārtha Nirūpaṇa*, Delhi: University of Delhi, pp. 608-30.

substances. These atoms are ideals, representing unities of fundamental attributes. In this sense, they are quite similar to the concept of such elementary particles of modern physics which are proposed on theoretical grounds.[23]

Theory of Chemical Action

Theory of chemical action or *pāka* is another significant feature of Vaiśeṣika which can be taken to be the earliest available theoretical reference in the field of chemistry in India. *Pāka* is defined as application of external heat[24] which results in the change of colour, taste, smell and touch pertaining to earth. This assertion is based on the notion that the four properties mentioned above are basically of two types: eternal and non-eternal; those inhering in eternal substrata are held to be eternal while those pertaining to non-eternal objects are non-eternal. But there is an exception in case of earth because these four properties abiding in earth are always held to be non-eternal and prone to change due to *pāka*. In other words, when heat is applied to any earthly object, then the colour, taste, smell and touch of that object are changed for ever[25] while in case of other material substances, this change is only temporary.

There are a lot of details available in the Vaiśeṣika texts regarding the process of chemical action and the different ideas about time-limit pertaining to it.[26] Although it is not relevant to refer to all of them here, yet one cannot avoid mentioning

23. Subhash Kak, "Indian Physics: Outlines of Early History", p. 12 (arXiv: physics/031000Iv1[physics. hist-ph] 30 Sep. 2003)

24. *pāko nāma vijātīyastejaúsaṁyogaḥ* — *Nyāyabodhinī* on *Tarkasaṁgraha*, pp. 17-18.

25. Vide Shashiprabha Kumar, 1991, "Nyāyavaiśeṣikayoḥ guṇānāṁ pāka-prakriyā", in *Sāgarikā*, 29.1, Varanasi: Sagarika Samiti, pp. 53-62.

26. *Kaṇādarahasyam*, p. 61; for explanation, see Umesh Mishra, 2006, *Nyāya-Vaiśeṣika*: *Conception of Matter in Indian Philosophy*, Delhi: Bharatiya Kala Prakashan, rev. edn., pp. 73-80.

that the Vaiśeṣika posits *pilupākavāda* (theory of chemical action in the atoms) while the Naiyāyika rejects this and proposes *piharapākavāda* (theory of chemical action in the whole as an object).[27] To explain these two theories through the popular example of pot, it might be said that Vaiśeṣika upholds successive baking of each atom in a pot, while the Nyāya propounds simultaneous baking of the complete finished product, i.e. the pot. It will not be out of place to add here that the modern method of heating through microwaves clearly proves that the Vaiśeṣika standpoint is more scientific.

Theory of Causation

Moreover, the Vaiśeṣika theory of causation in general and of homogeneous material causes in particular, besides the creation of a new whole which is different than its parts, are also worth examining from scientific point of view. In the history of Indian thought, Kaṇāda is the first to have defined and described the relation between cause and effect.[28] He explains the visible world by accepting and stating the basic notion of causality, i.e. the cause is a factor which points towards a substance, either because the effect inheres in it or because there is a connection which is operating from its base.[29] More notably, cause and effect are both having an individual identity in Kaṇāda's system as against the Sāṁkhya doctrine of shared identity between the cause and the effect. According to Kaṇāda and his followers, both the cause and effect, although two different entities, are inseparably tied through

27. *Upaskāra*, p. 371.

28. Satyaprakash Saraswati, 1988, *Prācīna Bhārata ke Vaijñānika Karṇadhāra*, Hindi Translation of *Founders of Sciences in Ancient India*, New Delhi: Pustakayan, rpt., p. 281.

29 (i) *iṣṭāniṣṭakāraṇaviśeṣādvirodhācca mithaḥ sukhaduḥkhayorthāntarabhāvaḥ* — *VS*, 10.1.1

(ii) *saṁśayanirṇayāntarābhāvaśca jñānāntaratve hetuḥ* — *VS*, 10.1.2

(iii) *tayorniṣpattiḥ pratyakṣalaiṅgikābhyām* — *VS*, 10.1.3

a unique relation termed *samavāya*.[30] It is due to this distinct theory of causation, that in Vaiśeṣika view each product is held to be a new beginning (*ārambha*) because it was not existing before its birth and will also not be present after its destruction. But the causal factors, being different than the effect, were there and will be there.

> It is a point to note that Kaṇāda is not dealing with causality in the empirical or practical sense, i.e. the sense in which the term is taken to connote mainly the process by which one thing or condition, say "a" is said to bring about an event or a condition, say "b". He treats of causality as a phenomenal feature of the world existence.[31]

Details regarding the chemical processes involved in production of a new whole (*avayavī*) from its constituent parts (*avayava*) also need to be discussed in scientific manner, which is well demonstrated by a modern exponent of Vaiśeṣika.[32]

Theory of Metals

The Vaiśeṣika view of holding the metals like gold, glass, silver and iron, etc. which are produced out of mines (*ākaraja*), as fiery objects (*taijasa*) is also very important in this regard. Significantly, fire according to Vaiśeṣika implies both heat and light. Kaṇāda has not explicitly mentioned anything about the classification of metals under fire (*tejas*), he has just stated that the metals like tin, glass, iron, silver and gold, etc. share a common feature with water, i.e. instrumental liquidity (*naimittika dravatva*).[33] But his followers have argued that the metals are fiery objects.

---

30. *ihedamiti yataḥ kāryakāraṇayoḥ sa samavāyaḥ* — *VS*, 7.2.26
31. Veena S. Gajendragadkar, 1988, *Kaṇāda's Doctrine of the Padārthas*, Delhi: Sri Satguru Publication, p. 71.
32. Vide Biswambhar Pahi, 2000, *Vaiśeṣika Padārtha — Vyavasthā kā Paddhatimūlaka Vimarśa*, Jaipur: Philosophy Department Rajasthan University, pp. 189-200.
33. *trapusīsa-loha-rajata-suvarṇānāmagnisaṁyogāddravatvamadbhiḥ sāmānyam* — *VS*, 2.1.7

There are two clear-cut views in this case — the first view is held by the opponents of Vaiśeṣika who assert that the metals like gold, etc. are earthly objects.[34] According to the second view, which has been put forth by several Vaiśeṣika scholars, metals like gold, etc. have to be taken as fiery products since their liquidity (*dravatva*) is not extinct even after application of excessive heat, while it is not so in case of other substances like earth and water, etc.[35] Therefore, the Vaiśeṣika argues, that metals like gold, etc. cannot be classified under earthly objects. The metals like gold, etc. cannot be covered under watery objects as well, since water has natural liquidity while the liquidity in metals is only instrumental. Besides, the luminous white colour (*bhāsvara śukla rūpa*) ascribed to fiery metals is also unique, for the earth and water have non-luminous white colour (*abhāsvara śukla rūpa*). Moreover, the metals can also not be classified under air because air does not have colour at all while the metals have luminous white colour. The fifth material substance ether (*ākāśa*), has neither colour nor liquidity, so gold, etc. cannot be clubbed with it too.[36]

In this way, through the process of elimination as mentioned above, one is left with no choice but to accept that the metals are fiery objects. It needs to be noted here that several objections have been raised against the Vaiśeṣika view of holding metals as fiery objects from within the tradition of Indian philosophy, but various Vaiśeṣika scholars have not only tried to respond to those but have also stood by their point of view on the basis of different logical arguments.[37]

However, there seems to be at least one empirically proved

---

34. *Kaṇādarahasyam*, p. 20.
35. *Upaskāra*, p. 120.
36. *Vivṛtti* on *Nyāyalilāvatīprakāśa* , p. 166.
37. Vide Shashiprabha Kumar, 1993, *Vaiśeṣikanaye Suvarṇādidhātūnāṁ Taijasatva-Siddhiḥ*, *Viśva-Saṁskṛtam*, 30.2, *Journal of VVRI*, Hoshiarpur, pp. 49-54.

factor which supports the Vaiśeṣika standpoint in this regard. Accordingly since metals are certainly known to be good conductors of heat, so they must belong to the same type of substance which is inherently associated with heat. Therefore, the Vaiśeṣika view of holding the metals as fiery products, stands vindicated from common experience as well.

Theory of Motion

The Vaiśeṣika exposition of motion (*karma*) is quite remarkable in the sense that it evinces a subtle analysis of numerous physical activities rather than discussing it as an act, process or verb (*kriyā*) as has been done in several other schools of Indian philosophy. Accordingly the Vaiśeṣika source texts explain a variety of physical motion such as throwing upwards (*utkṣepaṇa*), throwing downwards (*avakṣepaṇa*), contraction (*ākuñcana*), expansion (*prasāraṇa*) and movement (*gamana*) besides flowing of water (*syandana*) and falling down of weighty objects (*patana*), evacuation (*recana*), flaming up (*ūrdhvajvalana*) and curvilinear motion (*tiryaggamana*), etc. All these types of activity pertain to five material substances (*mūrta dravyas*) only, i.e. earth (*pṛthivī*), water (*jala*), fire (*tejas*), air (*vāyu*) and mind (*manas*) according to Vaiśeṣika.[38]

In this context it is interesting to note that motion, invariably produced in any one of the five material substances, is devoid of properties in itself, but is an independent cause in producing two properties, i.e. conjunction and disjunction according to Kaṇāda.[39] Praśastapāda, the earliest available commentator of Kaṇāda, while explaining the *Vaiśeṣika-Sūtras*, has listed several properties as different factors which cause various types of motions in five material substances.[40] Prof.

38. *PPB*, pp. 56-57.

39. *ekadravyamaguṇaṁ saṁyogavibhāgeṣvanapekṣakāraṇamiti karmalakṣaṇam* — *VS*, 1.1.17

40. For details, see *PPB*, pp. 697-98.

B.N. Seal has classified all these factors first into two groups such as mentioned below:

(A) Motions caused without any contact with material substances (*saṁyogājanya karma*)

(B) Motions produced by contact with matter (*saṁyogajanya karma*)[41]

Subsequently, various types of motions and their causes have been discussed by him under these two heads as follows:

(A) Motions caused without any contact with material substances:

(i) Motion caused through effort (*prayatna*) — In this case, activities like moving of hand for picking up food or for kicking an enemy are cited[42] in which there is no contact with matter at the level of production of the said activity.

(ii) Motion caused through weight (*gurutva*) — When a fruit falls from the tree due to its own weight and without any contact with matter, then the action would be covered under this group. In fact, contact with matter in such cases acts as a deterrent to falling, e.g. the contact of a bird's nest with the tree does not let it fall, even though it is prone to fall due to its weight.[43]

(iii) Motion caused through flowing of liquids (*dravatva*) — Liquids like water, etc. always flow downwards due to their natural and inherent liquidity.[44]

---

41. B.N. Seal, 1915, *Positive Sciences of the Ancient Hindus*, Delhi: Motilal Banarsidass, rpt. 1985, 1991, pp. 131-36.

42. *ātmasaṁyogaprayatnābhyāṁ haste karma* — *VS*, 5.1.1 and *PPB* with *Nyāyakandalī*, pp. 713-14.

43. *saṁyogābhāve gurutvāt patanam* — *VS*, 5.1.7 and *Upaskāra* on the above, pp. 299-300.

44. *dravatvāt syandanam* — *VS*, 5.2.4 and *Upaskāra* on the above, p. 311.

(iv) Motion caused through unseen factors (*adṛṣṭa*)[45] — There are several such motions which arise without any visible and explicable contact with matter; all these movements have been classified under this head. Initial motion in atoms, flowing upwards of fire, curvilinear motion of air, moving upwards of water in trees, etc. are all enumerated under this group.[46] Moreover, mental activities like outgoing (*upasarpaṇa*) and incoming (*apasarpaṇa*) which take place after the death of physical body, are also explained by Praśastapāda[47] as occurring due to *adṛṣṭa* (unseen moral factors).

Interestingly, Praśastapāda has also talked of mental activities pertaining to *yogins*, who have acquired special capabilities to transmit their mental faculties to the desired place and call these back as and when they wish to do so. Likewise, the initial mental activity for getting associated with a new physical body in the beginning of creation is also said to be caused by merit (*dharma*).[48]

(B) Motions caused through contact with matter:

(i) Motions caused through pressure (*nodana*): Pressure (*nodana*) is a specific type of continued

45. It needs to be clarified here that all those activities which are ascribed to *adṛṣṭa* by Kaṇāda, are in fact purely physical activities which were unexplainable up to his time. Later interpreters have, however, suggested moral implications of the term *adṛṣṭa* which do not seem correct. For details of changes in the meaning of *adṛṣṭa* from literal to moral aspects in the history of Vaiśeṣika, see Shashiprabha Kumar, 1999, "Nature and Role of Adṛṣṭa in Vaiśeṣika Philosophy", in *Facets of Indian Philosophical Thought*, Delhi: Vidyanidhi Prakashan, pp. 115-23.

46. *VS*, 5.2.13, 5.2.2, 5.2.5-7.

47. *PPB*, pp. 736-38.

48. In later Vaiśeṣika treatises, merit (*dharma*) and demerit (*adharma*) were equated with *adṛṣṭa*. Vide *Kārikāvalī* of Viśvanātha: *dharmādharmavadṛṣṭaṁ syāt*, verse 161.

contact which is either created out of a collection of these four properties — weight, liquidity, velocity and effort or out of any one, two or three of these. The motion in four material substances is generated through pressure only; e.g. the motion of an object pushed or pulled by hand, the motion of mud below the heavy stones, the motion of arrow due to the continued pressure exerted by the bow-string, the motion of bow-string due to the continued pressure of elasticity of the bow before it receives its original shape, the motion of clouds, of volumes of dust, of balloons filled with gas,[49] sailing vessels and other vehicles under the impelling force of the wind, etc. come under this category.[50]

(ii) Motions produced out of impact (*abhighāta*): Impact (*abhighāta*) is also a specific sort of contact which causes disjunctive motion in material substances through its velocity (*vega*); e.g. when a stone falls against a hard object, the potter's stick strikes the wheel or the mortar strikes against the pestle.[51] It is noteworthy here that impact (*abhighāta*) presupposes instantaneous disjunction, because if the contact is continued, then it will result into *nodana*.[52] However, in some cases, disjunction might occur after continued pressure (i.e. a rebound).

49. *Kiraṇāvalī*, p. 57. ***Note***: Prof. Seal has pointed out here that Udayana (CE 970) was well aware of the balloons filled with gas since he has mentioned these as *dhūmāpūritacarmapi aka* while referring to the weight of air.

50. *PPB*, pp. 725-26.

51. *Nyāyakandalī*, p. 727.

52. Thus while pressure is a contact that persists, impact is an instantaneous contact which is immediately followed by disjunction. Sadanand Bhaduri, 1975, *Studies in Nyāya-Vaiśeṣika Metaphysics*, Poona: Bhandarkar Oriental Research Institute, rpt., p. 122.

(iii) Motions caused by elasticity (*sthitisthāpakatā*): When any material object, after being pulled for a while, exercises a moving force due to its elasticity and regains its original form, then the motion in that object is caused by its elasticity. For example the motion of a bow-string due to the force exercised by the bent bow is caused due to elasticity of the same.[53]

(iv) Motions caused through velocity or momentum (*vega*): This includes all such motions which take place due to contact with a body that itself is in contact with something else having impressed motion. In other words, such motion arises in an object which is in contact with any thing already moving with velocity; momentum is the outcome of transmission of that moving object. This momentum in turn causes movement in the former object. For example when an arrow starts moving, then its subsequent motion is caused due to the momentum already present in the moving arrow and the same continues till such time that it falls due to loss of momentum and weight.[54]

It is noteworthy here that motion is held to be momentary in Vaiśeṣika[55] and it cannot be caused by another motion according to Vaiśeṣika theory, hence the acceptance of velocity (*vega*) as one of the factors for successive movement in an

53. *PPB*, pp. 722-23.

54. *nodanādādyamiṣoḥ karma tatkarmakāritācca saṁskārāduttaraṁ tathottaramuttaraṁ ca*. *VS*, 5.1.17 and *Upaskāra* on the above, p. 306.

55. It is here that the Vaiśeṣika scholars discuss the idea of momentariness (*kṣaṇikatva*) of motion, since motion is momentary by definition. However, the Vaiśeṣika idea of momentariness does not amount to the Buddhist identification of existence and flux, since it covers phases of several moments. Vide *PPB* and *Nyāyakandalī*, p. 65.

object is essential for explaining the continuity of motion. The Vaiśeṣika idea of accepting and explaining velocity (*vega*) as one of the significant causal factors for motions not only bears remarkable resemblance with Newtonian laws of motion[56] but also proves the scientific attitude of Vaiśeṣika scholars.[57]

In addition to the above eight exclusive factors responsible for causing motion, Praśastapāda has mentioned that in some cases, two of the above factors are also seen operating jointly as follows:

(i) Pressure (*nodana*) and velocity (*vega*)

(ii) Impact (*abhighāta*) and velocity (*vega*)

(iii) Pressure (*nodana*) and impact (*abhighāta*)

(iv) Weight (*gurutva*) and velocity (*vega*)[58]

Besides discussing these causal factors for producing various types of motion, Praśastapāda has also explained some other specific motions such as vertical motion (*saṁmūrchana*),[59] vibratory motion (*kampana*),[60] rotatory motion (*bhramaṇa*),[61] current motion or downward flow (*syandana*), upward motion (*ārohaṇa*)[62] and capillary motion (*abhisarpaṇa*).[63]

---

56. N.G. Dongre, 1979, *Vaiśeṣikasiddhāntanāṁ Gaṇitīyapaddhatyā Vimarśaḥ*, Sampurnanand Sanskrit University, Varanasi, pp. 82-83.
57. "They had a good grasp of the concept of force and were able to indicate the resultant motion when a particle was subjected to a number of forces or motions. The direction depends on the fact that pressure impact is made in the original or opposite direction". S. Radhakrishnan, 1967, *History of Philosophy: Eastern & Western*, vol. I, George Allen & Unwin, London, 3rd impression, pp. 460-61
58. *PPB* with *Nyāyakandalī*, pp. 714-24.
59. *PPB* with *Nyāyakandalī*, p. 119.
60. *Karīkāvalī*, 157-59.
61. *Nyāyakandalī*, p. 732.
62. *VS*, 5.2.4-5 and *Upaskāra* on the above, pp. 311-12.
63. *Nyāyamanjarī*, vol. II, p. 51.

Theory of Dream

As regards the Vaiśeṣika theory of cognitive processes, it is usually discussed under the property called *buddhi* or *jñāna* in Vaiśeṣika treatises. The classification of *buddhi* as described by Praśastapāda[64] is given as follows:

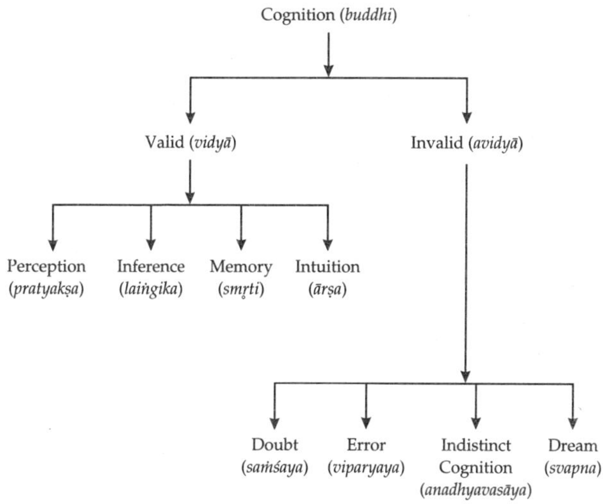

Details of all the processes listed above cannot be provided here, but a special mention must be made of the dream (*svapna*) cognition which is more relevant with our present discussion. Accordingly, dream (*svapna*) is still being widely discussed in applied psychology and might need further explanation in future as well, since it is a very complex psycho-physical experience which evades explanation. As delineated by Kaṇāda and Praśastapāda, dream is a mental cognition which arises out of previous impressions but is different from memory.[65]

Dream is classified under invalid cognition, since there is no real ground for its existence in the external world and there is no actual contact between the sense organ and the object.

64. *PPB*, pp. 411 and 441.

65. For details see Shashiprabha Kumar, 1999, "Vaiśeṣika Manodarśana meiṅ Svapna-Siddhānta" in *Vaiśeṣika Darśana Pariśīlana*, Delhi: Vidyanidhi Prakashan, pp. 64-75.

Dream is a unique mental experience which takes place when all the physical senses are inactive but mind is active. It has psycho-physical aspects as will be clear from glancing through the triple factors which cause three types of dreams as follows:

(i) Psychological (*saṁskārapāṭavajanya*) — Which occurs due to the continuity of mental impressions gathered during daytime. It is due to the freshness and intensity of these impressions that such dreams arise.

(ii) Physiological (*dhātudoṣajanya*) — Which appears on account of physiological imbalances. This type of dreams are actually based on the triple interpretation of essential constituents (*dhātu*) within the human organism which are discussed in Āyurveda.

(iii) Inexplicable (*adṛṣṭajanya*)[66] — There are some such dreams which have no direct link with the physical or mental experiences of the person involved but there is no denying the fact that one does have dreams of this type too. So all such dreams can be covered under this head.

Interestingly, the Vaiśeṣika scholars have also talked of dream-end (*svapnāntika*)[67] cognition which refers to the experience of "dream within a dream". Since such an experience occurs towards the end of dream, hence this is termed as "dream-end" (*svapnāntika*). However, there is a difference of opinion regarding its classification; Kaṇāda has not specified its grouping different than the dream (*svapna*) while Praśastapāda and his commentators seem to be putting it under memory.[68] Vyomaśiva has clearly stated that dream is a mental experience which takes place after the external sense organs have stopped operating, but dream-end is a form of recollection.[69]

66. *PPB*, p. 439.

67. *svapnāntikam* — *VS*, 9.2.8.

68. *PPB* with *Nyāyakandalī*, pp. 441-42.

69. *Vyomavatī*, p. 549.

Conclusion

What follows from the above brief delineation is that the scientific approach of Vaiśeṣika system is well reflected in its enunciation of physical phenomena as well as in exposition of mental processes.

In short, it would not be incorrect to say that the Vaiśeṣika texts provide a rich repository of scientific insights in diverse areas such as physics, metaphysics, chemistry, physiology, psychology and metallurgy, etc. Vaiśeṣika views on the nature of sound, number and negation are also worth exploring in this regard which have not been discussed herein. In the traditional circles of Indian philosophy, the system of Vaiśeṣika might not have been respected and eminent scholars like Śaṅkara have not only characterised it as not committed to Veda (*vedavirodhī*) but have also branded it as semi-nihilist (*ardha-vaināśika*), but in the modern rationalistic and scientific approaches, it is bound to gain ground as can be evinced from the growing worldwide interest in the study of Vaiśeṣika. The same has been echoed by a modern Vaiśeṣika scholar in the following manner:

> On the other hand, the realistic, physicalistic, and analytical orientation of the Vaiśeṣika and allied Nyāya system has attracted the special attention of modern Western and Indian scholars. It has been claimed as an exemplary expression of the non-mystical, secular, analytical, and scientific potential of Indian thought. James R. Ballantyne was convinced that the rediscovery and propagation of Nyāya-Vaiśeṣika in India would provide a basis and framework for the introduction of modern Western science.[70]

---

70. Wilhelm Halbfass, 1993, *On Being and What There Is: Classical Vaiśeṣika and the History of Indian Ontology*, Delhi: Sri Satguru Publication, I[st] Indian edn., pp. 81.

References

Vaiśeṣikadarśana with *Praśastapādabhāṣya* with *Prakāśikā*, Hindi Commentary by Acharya Dhundhiraja Shastri and edited with introduction and Hindi-Translation of the *Vaiśeṣika-Sūtra* by Sri Narayan Mishra Varanasi: Chaukhambha Sanskrit Sansthan, 2002 (4th edn.).

Praśastapādabhāṣya with *Nyāyakandalī*, Edited and translated in Hindi by Pt. Durgadhar Jha, Varanasi: Sampurnanand Sanskrit University, 1997.

Kiraṇāvalī of Udayanācārya, ed. by Jitendra S. Jetly, Baroda: Oriental Institute, 1971.

Praśastapādabhāṣyam by Praśastadevācārya with Commentaries (up to *Dravya*) *Sūkti* by Jagadīśa Tarkālaṁkāra *Setu* by Padmanābha Miśra *Vyomavatī* by Vyomaśivācārya, ed. by Gopinatha Kaviraj and Dhundhiraja Shastri, Banaras: Chowkhamba Sanskrit Series Office, 1930.

Vaiśeṣikasūtropaskāra of Śaṅkara Miśra with the *Prakāśikā*, Hindi Commentary by Dhundhiraja Shastri, ed. Sri Narayan Mishra, Varanasi: Chaukhamba Sanskrit Sansthan, 2002 (2nd edn).

Nyāyalīlāvatī with *Kaṇṭhābharaṇa* of Śaṅkara Miśra, *Prakāśa* of Vardhamana Upadhyaya and *Vivṛtti* of Bhagiratha Thakur, ed. Harihara Shastri, Varanasi: Chowkhamba Sanskrit Series Office, 1991 (2nd edn.).

Nyāyamañjarī of Jayanta Bhaṭṭa, ed. Surya Narayan Shukla, Varanasi: Chowkhamba Sanskrit Series Office, 1971.

Kaṇādarahasyam of Śaṅkara Miśra, ed. by V.P. Dwivedi, Banaras: Chowkhamba Sanskrit Series, XIV, 1917.

Kārikāvalī of Viśvanātha with the Commentaries *Muktāvalī*, *Dinkarī*, *Rāmrudrī*, ed. by Atmaram Narayan Jere, Varanasi: Krishnadas Academy, 1988 (rpt.).

Saptapadārthī of Śivāditya, ed. and translated in English by D. Gurumurti, Madras: Theosophical Publishing House, Adyar, 1932.

Tārkikarakṣā of Varadarāja, ed. by Vindhyesvari Prasad Dwivedi, Banaras, 1903.

Tarka-Saṁgraha of Annambhaṭṭa with the Author's Own *Dīpikā* and Govardhana's *Nyāya-Bodhinī*, ed. Yashwant Vasudeva Athalye and Mahadeva Rajaram Bodas, Poona: Bhandarkar Oriental Research Institute, 1974 (2nd edn.).

4

Fourfold Description of Sound
A Vedic View

M.G. Prasad

Acoustics is a field of science and arts dealing with generation, transmission and reception of sound. Acoustics has received a very high importance in Vedas and Vedic literature. The classical literature of Hinduism or Sanātana Dharma is rooted in Vedas. Sound in modern scientific terms refers to the pressure waves in air received by human beings in the audible range of generally 20 cycles/sec to 20000 cycles/sec. However the infra- and ultra-sound refers to the frequencies below and above the audible range respectively. In Vedas and Vedic literature several terms such as *nāda* and *śabda* are used. These terms cannot be translated directly as sound as they refer to vibrations including both physical and spiritual aspects. However the *nāda* and *śabda* include the human auditory range sound. In particular human speech, vocal and instrumental music that lie within the audio range have been discussed significantly in Vedas and Vedic literature. This paper presents a Vedic view of the fourfold description of sound expressed in the form of human speech. Also the paper includes discussions of Vedic view in relation to the views of modern acoustics.

Introduction

Veda refers to the body of knowledge that deals with physical, psychological and spiritual aspects of life and cosmos. The word *veda* etymologically relates to the verbal root *vid* meaning "to know". However it is interesting to note that the Vedas also refer *mantras* meaning "chants that convey knowledge". The Vedic chants have acoustical characteristics such as phonetics, tonal quality, length measure, effort, etc. These Vedic chants are orally transmitted with precision from master to disciples over thousands of years. It is well known that the precision in the oral transmission of chants, music and speech cannot be surpassed through writing. Thus acoustics of human speech is very important. Also it is well known that the faculty of speech is revered as Goddess of Speech (Vāgdevī). The *mantra* from the *Yajurveda* says:

> *devīṁ vācamajanayanta devāḥ, tam viśvarūpaḥ paśavo vadanti*
> *sā no mandreśamūrjam duhānā dhenurvagasmānupasśtutaitu*

The meaning of this *mantra* could be summarised as Creator Lord Brahmā and other *devatā* brought forth Vāgdevī (Goddess of Speech) to manifest communication. All beings use their abilities to communicate through Vāgdevī. May Vāgdevī like an all-desire fulfilling cow bless us with food, strength and faculty of speech.

It is quite difficult to precisely define and describe the origin of language and human speech. However it is known that Vedic chants and music in India have existed from very ancient times. Sir C.V. Raman in his article titled "The Acoustical Knowledge of Ancient Hindus"[1] says that:

> It would form a fascinating chapter of history to try and trace the gradual development of musical instruments and musical knowledge, from the rhythmic chanting of the *Ṛgveda* in the ancient home of the Āryan race to the Indian music of the present day.

Speech Production: A Modern View

A question of scientific importance is about the origin of sound. This question, to begin with, can be addressed in reference to human speech. There is a large amount of scientific literature

1. C.V. Raman, 1922, *The Acoustical Knowledge of the Ancient Hindus, Āśutoṣa Mukherjī Silver Jubilee Volume*, vol. 2, pp. 179-85.

on human speech and music.[2] In a paper by Honda[3] four processes are recognised in production of speech. They are:

(a) Language processing, in which the content of an utterance is converted into phonemic symbols in the brain's language centre;

(b) Generation of motor commands to the vocal organs in the brain's motor centre;

(c) Articulatory movement for the production of speech by the vocal organs based on these motor commands;

(d) Emission of air sent from the lungs in the form of speech.

It is seen from the above description that the emphasis is on neurological and physiological basis. It is the intention of this paper to look into the importance of speech sounds and their production from Vedic view.

Importance of Sound Field: Vedic View

The Vedas and Vedic literature are the classical source literature for Sanātana Dharma which includes sciences and arts of various aspects of life. It is well known that the Vedas are collections of *mantras*, which have been orally transmitted through the ages. It is interesting to note that UNESCO has also recognised the tradition of the oral transmission of Veda *mantras* as below:

> Expressed in the Vedic language, which is derived from classical Sanskrit, the verses of the Vedas were traditionally chanted during sacred rituals and recited daily in Vedic communities. The value of this tradition lies not only in the rich content of its oral literature but also in the ingenious

2. (a) M. Honda, 2003, "Human Speech Production Mechanisms", *NIT Technical Review*, vol. 1(2), pp. 24-29.

   (b) A.D. Patel, 2008, *Music, Language and Brain,* New York: Oxford University Press.

3. M. Honda, 2003, op. cit., pp. 24-29.

> techniques employed by the brāhmaṇa priests in preserving the texts intact over thousands of years. To ensure that the sound of each word remains unaltered, practitioners are taught from childhood complex recitation techniques that are based on tonal accents, a unique manner of pronouncing each letter and specific speech combinations. — *UNESCO Proclamation 2003* (*Intangible Cultural Heritage* – ICH)

The precision in oral transmission is directly related to acoustics. The six acoustical factors involved in precise Vedic chanting as per *Taittirīya Upaniṣad* are pronunciation of letters (*varṇaḥ*), their pitch (*svaraḥ*), the timing or duration (*mātrā*), the force (*balam*), melodious articulation (*sāma*) and combined sound effects (*santānaḥ*).

The sound and production of sound from human beings as speech and music are part of the *nāda* and *śabda*, which can be called a *sound field*. The various manifestations of sound in the world are generated from sound field. The sound field as *nāda* and *śabda* has received utmost importance in Vedic literature. Śāraṅgadeva, the author of *Saṅgīta Ratnākara*, a treatise on classical Indian music says:

> *nādena vyajyate varṇaha padam varṇāt padadvacaha* |
> *vacaso vyavahāroyam nādādhīnamato jagat* | |

The alphabets have manifested from the *nāda* (sound field). The words are made up from alphabets and from the words speech come out. Worldly transactions are carried out through the speech, and thus, the world is dependent on *nāda*.

In a similar way, in the words of a Yogī Sri Ranga Sadguru,

> Just as from seed comes naturally sprout, branch, leaf, flower, unripe fruit and full fruit likewise starting from spiritual light as seed, the *nāda*, the *svara* and *akṣaras* have developed into the various aspects of knowledge[4]

4. Śrīraṅga Sadguru, 1983-2008 (19 vols.), *Amaravaṇī: Lectures on Sanatana Dharma*, Mysore: Aṣṭāsṅga Yoga Vijñāna Mandiram.

The sound field as *śabda* and *nāda* takes us to the origins of the universe. The *Amṛta Bindu Upaniṣad*,[5] one of the minor Upaniṣads, the following verse says:

> *dve brahmaṇi veditavye śabda brahma paraṁ ca yat* |
> *śabda brahmaṇi niṣṇātaḥ paraṁ brahmādhigacchati* | |

> Two *Brahmans* to be realised, namely *Śabda* and *Param Brahman*. One who has realised and is well versed in *Śabda Brahman* will realise *Param Brahman*.

Thus it is seen that the knowledge of sound field as *śabda* and *nāda* is a prerequisite to the spiritual fulfilment in life. Also the sound field as the revered *Nāda Brahman* in addition to its being all-pervasive in the world is also source of Bliss. Śāraṅgadeva, the author of *Saṅgīta Ratnākara*,[6] says:

> *caitanyam sarva bhūtānām vivṛtam jagadātmanā* |
> *nāda brahma tadānandam advitīyamupāsmahe* | |

> We worship the *Nāda Brahman* second to none which is blissful and is in all beings as consciousness has manifested itself as universe.

The scriptures also refer to two types of *nāda*, namely *anāhata* and *āhata*. The *anāhata* refers to self-existing unproduced sound field without any vibratory cause and is experienced only in spiritually focused deepest states of Yoga. The *āhata* refers to the sound field produced by vibratory cause. The discussion in this paper deals with produced sound field, namely *āhata nāda*.

---

5. K.N. Aiyar, 1997, *Thirty Minor Upaniṣads* (English Translation and Sanskrit text), Delhi: Parimal Publications.
6. R.K. Shringy, and P.L. Sharma, 1978, *Saṅgīta Ratnākara of Śāraṅgadeva* (text and translations), vol. I, Delhi: Motilal Banarsidass.

Four Stages of Speech Sounds: Vedic view

An important hymn from *Taittirīya Brāhmaṇa*[7] says:

> *catvāri vāk parimitā padāni tāni vidur brāhmaṇā ye maniṣiṇaḥ* |
> *guhā trīṇi nihitā neṅgayanti turīyam vāco manuṣyā vadanti* | |

> Speech or word has four stages which are known by brāhmaṇa. Who has control over mind is well versed in this field. The first three stages are concealed in the cave within while only the fourth stage comes out as utterances by human beings.

According to a view of philosophers of grammar[8] the four stages are referred as (a) *parā,* (b) *paśyantī,* (c) *madhyamā,* (d) *vaikharī.*

Also in addition scriptures refer to *dhvani*[9] which would mean as voice and is related to speech sounds. The author Śāraṅgadeva in *Saṅgīta Ratnākara*[10] says:

> *ātmā vivakṣamāṇayoam manaha prerayate, manaha* |
> *dehastham vahnimahanti sa prerayati mārutam* | |
> *brahma granthisthitaha so'etha krāmadūrdhvapathe caran* |
> *nābhihṛtkaṇṭhasyiśeśvāvirbhāvayati dhvanim* | |

> The human being impels the mind and the mind activates the internal power source in the body which in turn stimulates the vital force. The vital force stationed around the root of the navel, rising upwards gradually through the heart, and the cerebrum and the cavity of the mouth as it (the vital force) passes through them to manifest as *dhvani.*

---

7. (a) N.S. Ramabhadrācārya, 2004, *Udaka Śānti Mantrāḥ* (Sanskrit and Translations in Kannada), Mysore: Sri Sanatana Bharati Pathasala.

   (b) Rama Nath Sharma, 2007, *Vāk, Śabda-tattva and Veda,* Ujjain: World Vedic Conference.

8. Rama Nath Sharma, 2007, op. cit.
9. S. Mukhopadhyaya, 1979, "Importance of Sound in the Tradition of Vedic Chanting", in *Dhvani,* ed. S.C. Malik, New Delhi: Indira Gandhi National Centre of Arts and D.K. Printworld.
10. R.K. Shringy, and P.L. Sharma, 1978, op. cit.

In this process also we can see clearly the four stages for production of sound or *dhvani* as:

(a) the person's impelling of the mind;

(b) the mind activating the vital force and getting ready to move;

(c) the vital force moving and rising up through the navel, heart, throat, cerebrum and mouth cavity;

(d) *dhvani* produced and travelling to receiver.

Parā, Paśyantī, Madhyamā and Vaikharī

It is to be noted that in Vedic view the origin of the verbal speech in human beings is traced to the location of the *mūlādhāra cakra,* which is placed in the coccygeal triangle where the spinal cord ends. This energy centre or *cakra* is of primary importance as it supports the entire body organisation.[11] An important verse from the *Prapañcasāra Tantra* text says:

> *mūlādhārāt prathamamudito yastu bhāvaḥ parākhyaha* |
> *paścāt paśyantyathā hṛdayāgo buddhiyuṅgmadhyamākhyaha* | |
> *vaktre vaikharyatha rurudiṣorasya jantoḥ suṣumnā* |
> *baddhastasmādbhavati pavana prerito varṇa saṅghaḥ* | |

> The *parā* form of speech itself was unmanifest to begin with, then it manifested to become *paśyantī* as it moved up with vital force to the region of umbilicus or navel area. (Also simultaneously, the information at knowledge level is being converted into speech form.) The *paśyantī* form formed at the location of navel region rises up along with the vital force to the location of heart to become *madhyamā* along with the integration of the intellect to transform knowledge into speech form. Then *madhyamā* form at the heart rises along with the vital force to the location of throat and mouth

11. N.S. Ramabhadracharya, 2003, *Śrī Guru Hṛdaya Ārṣa-sāhitya and I,* (Sanskrit text and translations), Mysore: Aṣṭāṅga Yoga Vijñāna Mandiram.

> to manifest as *vaikharī* speech or verbal speech heard by the listener. Thus the articulated speech that contains alphabets (and words) have manifested through the movement of vital force through the central energy carrying channel termed as *suṣumnā* along the central axis of the spinal cord.

Thus in the above verse the generation of whole process of fourfold level of speech is well described from Vedic view.

There is a verse from a classical Hindu text, which specifically gives the locations in terms of *cakra*s to the fourfold level of speech in physiological terms in the human body similar to that as the previous verse. The verse says:

parā vaṅgmūlacakrasthā paśyantī nābhisaṁsthitā |
hr̥dgā tu madhyamā jñeyā vaikharī kaṇṭhadeśagā | |

> In the fourfold speech, the *parā* level speech manifests at the *mūlādhāra cakra* (coccygeal triangle), the *paśyantī* and *madhyamā* at the *anāhata cakra* (cardiac or heart region) and the *vaikharī* at the *viśuddhi cakra* (at the cavity of the throat).

It is seen that similar description is given in the *Bhāgavata Purāṇa*[12] in the verse 11.12.17.

In another verse, the psychological process of speech formation is given as:

vaikharī śabda niṣpattihi madhyamā smr̥ti gocarā |
dyotikārthasya paśyantī sūkṣmā brahmaiva kevalam | |

> In the fourfold levels of speech, the *vaikharī* form is the gross manifestation of words. The *madhyamā* form is at the memory level, the *paśyantī* form is at the level where the motive is seen and the *parā* form is at the subtle unmanifest as *Brahman*.

The above three important verses clearly show that the Vedic

12. G.L. Beck, 1993, *Sonic Theology*, USA: University of South Carolina Press.

view integrates both the physiological and psychological process involving the human body in explaining the fourfold speech (or sound field) formation through (a) *parā*, (b) *paśyantī*, (c) *madhyamā*, and (d) *vaikharī* levels.

The *vaikharī* level is the production of words and sentences which is physically energised by the vital force (in the form of exhalation and inhalation) at the throat and mouth. Prior to this is the *madhyamā* level drawing from memory (refers to vocabulary) and is physically energised by the vital force at the heart or cardiac region. Prior to this is the *paśyantī* level where the purpose of speech is seen by the speaker and is physically energised by the vital force at the region of navel area or region of umbilicus. Prior to this is the subtlest and unmanifest *parā* level and is physically exists as potential energy form at the base of the spine or the coccygeal triangle. It is interesting to observe that in production of speech and music, not just the contributions of the efforts of the mouth but of the whole body is involved.

Relevance of the Vedic Fourfold Model of Speech

A comparison of the modern acoustical view and the Vedic view of the production of the speech is considered. It is seen that the modern view considers the role of brain, lungs, and vocal organs such as throat and mouth. However, the Vedic view integrates through fourfold levels in both the physiological and the psychological aspects. The physiological aspects in Vedic view go to the base of the spinal cord as the source of physical energy which drives the vital force to rise up and activate the various components such as lungs, heart, throat and the mouth. The psychological aspects do bring the role of brain through memory and seeing the motive of speech. The Vedic view seems to be more comprehensive. In addition, the Vedic view of speech as a divine energy is important for the human beings in their spiritual developments.

The Vedic fourfold level speech production can be applied to any type of sound production. As an example, when one blows a conch-shell, the sound produced by shell (heard by the listener) is *vaikharī*. Prior to that the blowing of vibrating air-pulses to the conch-shell is the *madhyamā* level. Prior to that of blowing, the person blowing the conch-shell had decided to blow for a particular purpose which is seen by the person and is at the *paśyantī* level. Prior to that is the capability or potential of the person to blow a conch-shell, which is at unmanifest level being referred as *parā*. This Vedic fourfold approach for speech production can be adapted to any production of the sound either through direct vocal music or through an instrument.

Concluding Remarks

The Vedic fourfold approach to sound production provides a deep insight to understanding many phenomena. There are ongoing efforts to analogically view the creation of the universe through this Vedic fourfold view of production of sound field as (a) *parā,* (b) *paśyantī,* (c) *madhyamā,* and (d) *vaikharī.*[13] Another way of observation is from manifest stage to unmanifest stage, that is to say, from *vaikharī* to *parā*. The interesting feature of the *vaikharī, madhyamā, paśyantī* and *parā* (VMPP) model is that one can direct the efforts to understand the life and nature starting from *vaikharī* and proceed towards *parā*. It is like going from sound to deep silence. The important point to be noted is that there is continuous energy link through vital force from *parā* to *vaikharī*. This means that going in the reverse direction, one should be able to start from *vaikharī* and proceed towards the *parā*, which is spiritually at a very deep level.

Although *vaikharī* is the last quarter but one can experience *vaikharī* as it is fully in manifested form. It encompasses the

13. A. Dalela, 2008, *Vedic Creationism*, i Universe, Inc., New York, USA.

whole life in many of its expressions. The *vaikharī* level can be seen and experienced in its expressions such as speech, music, literature and spirituality.[14] We know that speech directly relates human transactions. Music is the human artistic expression.[15] The literature such as poetry, plays, come into life through sound. The spirituality expresses through rituals and meditation in which *mantras* are used.[16] It is known in the spiritual literature the importance of many types of Yoga are based on the sound field such as *mantra-yoga, nāda-yoga, japa-yoga, svara-yoga,*[17] etc. Thus further work is needed in both theoretical and practical understanding and applications of the sound field.[18] In this regard, fourfold Vedic model for sound field would be helpful.[19] Further understanding of the sound field would also be rewarding as it brings joy in one's life.

Acknowledgement

The author is grateful to Yogī His Holiness Sri Ranga Priya Swami of Aṣṭāṅga Yoga Vijñāna Mandiram, Bangalore, India, for his guidance and help. Also, the author gratefully thanks Dr. V. Sadagopan of New York, USA, for his help.

---

14. M.G. Prasad, 1998, *Garland: An Anthology of Hinduism,* New York: Association for the Advancement of Arts and Sciences from India, USA.
15. R. Varma, 2009, *Indian Music,* New Delhi: Sasta Sahitya Mandal.
16. (a) Sir J. Woodroffe, 1985, *Garland of Letters,* Ganesh and Company.

    (b) S.V. Chamu. 2003, *The Divine Dancer,* Mysore: Aṣṭāṅga Yoga Vijñāna Mandiram.
17. (a) Brahmavarchas, 1995, *Śrī Ram Ācārya Vāṅgmaya (Śabda-Brahma-Nāda-Brahma),* Mathura: Akhand Jyoti Samsthan.

    (b) Śivānanda, Svāmī, 1994, *Tantra Yoga, Nāda-Yoga* and *Kriyā-Yoga,* Divine Life Society, Uttar Pradesh.
18. B. Seal, 1985, *The Positive Sciences of the Ancient Hindus,* Delhi: Motilal Banarsidass.
19. A. Padoux (tr. Gontier, J.), 1990, *Vāk (The Concept of the Word in Selected Hindu Tantras),* New York: State University of New York Press.

5

Relevance of Vedic Physics in the Contemporary Society

Sadasiva Biswal

The Vedic literature is the primordial source of human culture and civilisation. The Vedic texts include the four Vedas, the Saṁhitās, the Brāhmaṇas, the Āraṇyakas, the Kalpasūtras, the Upaniṣads and the Vedāṅgas (the auxiliary treatises). Besides, religion, philosophy and mythology, the Vedic literature contains various branches of knowledge, such as Astronomy, Astrology, Cosmogony, Cosmology, Space-time Continuum, Mathematics, Chemistry, Physics, Meteorology, Seismology, Botany, Zoology, Geology, Medicine, Agriculture, Environmental Science and Ecology and above all elements of Science and Technology. Ostensibly, the Veda means "the knowledge *par excellence*". The Vedic bards were natural scientists. The significant aspects of science and scientific thoughts of the Vedic bards have been reflected in their hymns. After development of modern scientific theories and laws, it is observed that the speculations of the Vedic seers agree well with the present knowledge of scientific discoveries and inventions.

The Vedic deities symbolise the natural phenomena which the human beings of that period came across in their day-to-day life and had interactions with such events. Evidently, the modern scientific theories and practices have their roots in the most ancient tradition and culture. The present discourse is the outcome of the cultivation of new thoughts and insights in the study of Vedic literature in its microscopic character over a macroscopic system.

Introduction

THE word "Science" has its derivation in the Latin word

Scientia, meaning "to know". Knowledge obtained through science should not be solely factual but should advance understanding of phenomena. Obviously, any knowledge is not called science nowadays. Science means today what is called *systematised and formulated knowledge*. It has been classified according to certain principles. In order to understand these principles we have to remember that our universe consists of matter, energy and life. These are all collectively called the nature. So, natural philosophy or natural science concerns all the phenomena in nature, the phenomena being partly physical and partly biological. It is also noticed that science begins from philosophy and finally it ends in philosophy.

An enquiry into the origin and nature of ancient Indian materialism shows that the materialism has been blended with the philosophical thoughts. Oriental idealism is commonly contrasted with accidental materialism as if there were races of men either naturally materialistic or naturally idealistic. In fact, a survey of Indian philosophical system clearly points out that materialism was never unknown to our thinkers and "it undoubtedly had its day in India".[1] Indian philosophy has had an extremely long and complex development — much more complex and probably a longer history of continuous development than any other philosophical tradition. Thus, here we find varieties of thoughts — scepticism, materialism, pluralism, dualism, naturalism, along with idealism of different forms. Philosophy, being man's quest after the ultimate reality, every philosophical tradition, including that of India, is characterised by so many divergent viewpoints.[2] After surveying the history of materialism, Lange[3] declares

1. Radhakrishnan and Moore (eds.), *A Source Book of Indian Philosophy*, p. xxiii.
2. U. Gupta, *Materialism in the Vedas*, New Delhi: Classical Publishing Company, 1987, pp. 1-2.
3. Lange, *History of Materialism*.

that not only materialism is as old as philosophy, but the latter itself started its career as the former. Therefore, it would be correct to say that materialism is the oldest known philosophy. Russell[4] has identified this earliest philosophy with science.

Formerly, Vedic people, in the absence of acceptable explanations of phenomena, developed their own "informal science" based on magico-religious speculations which subsequently gave birth to true scientific observations in the theoretical and experimental perspectives. Our three R̥gvedic *sūktas*, viz. *pr̥thivī, agni* and *savitr̥* correspond to Matter, Energy and Life needed for the genesis and growth of any culture. The ancient Indian science sprang from the magico-religious phenomena narrated in our Vedic literature.

The Vedic concept of *r̥ta*, implying some consciousness of the law-bound universe and the belief in the magical efficacy of *mantra* and *yajña* as the means to have control over the courses of nature and even over gods, underlies an attitude of the Vedic Āryans akin to that of science. Frazer,[5] in fact, describes magic as the "next kin to science" because both have in common "the general assumption of a succession of events determined by law". Needham[6] points out that in its earliest phase, science was indistinguishable from magic; he calls the latter "pre-science" or "proto-science". According to him, this scientific phase of the "magico-scientific" tradition, which in its beginning "crystallised around the nucleus of primitive shamanism", has left its unmistakable imprint on the people not only in ancient and mediaeval periods but even after the birth of modern science.

The hypotheses and the speculations of the Vedic seers go a long way in exploring the scientific causes and effects created

4. Russell, *A History of Western Philosophy*, p. 21.
5. Frazer, quoted in *Vedic Science and Technology*, by S. Biswal and B.L. Ray, New Delhi: D.K.Printworld, 2009, p. viii, Preface.
6. J. Needham, *Science and Civilization in China*, London: Cambridge, 1956, vol. I, pp. 108-209.

every moment in nature. It is, thus, revealed that the modern scientific theories and practices have their bases in the most ancient tradition and culture.

Physics

Physics is that branch of science which deals with matter and energy in space and time. This component of science began with the very moment of the creation of the material world. The origin of existent from non-existent is the starting point of this material world. Physics embraces the subjects like Astronomy, Cosmogony, Cosmology, Space and Time Continuum and Relativity, etc. Here in this paper, we will focus our attention to these subjects in perspective of Vedic as well as modern scientific thoughts.

VEDIC THEORY OF CREATION

There are two theories of creation described in the Vedas, namely

- (i) Shell theory of creation
- (ii) Water theory of creation

According to the shell theory of creation, first, a cosmic nucleus (Prajāpati) was formed. Then, he created the material world, i.e. *khip teja, marut, vyoma* and *āpa*, or the earth, the fire, the air, the *antarikṣa* and the water. These are the five elements of the material world. After this, life came to exist. This cosmic nucleus is often named as *hiraṇyagarbha* or "the Golden Egg", which is considered as the source of all matter and energy. All discrete substances of the Universe were aggregated here in the Golden Egg with enormous energies. The primeval body of this cosmic nucleus was disintegrated by the supreme spirit (the desire for creation). This happened with a great explosion producing sound of a very high pitch. *Chāndogya Upaniṣad*[7] also says that the universe came to exist

7. *Chandogya Upanisad*, VI.32.3.

after dismemberment of the cosmic egg. This explosion that brought about the creation is nothing but the Indra–Vṛtra collision mentioned in the Veda. Indra–Vṛtra episode narrated in the Veda goes like this.

Vṛtra was a demon representing the great inertia of Newtonian mechanics. Vṛtra was opposing Indra to continue his (Indra's) reign. Here, Indra represents the force that can initiate motion. According to Newton's first law of motion, i.e. law of inertia, a body continues in its state of rest or uniform motion in a straight line unless an external unbalanced force acts on it to change its state of rest or uniform motion. That means inertia opposes the motion. If the applied force will be stronger, then the inertia will be broken and the motion will start. Actually this happened. Indra threw his *vajra* on the demon Vṛtra and killed him. At this time, a great explosion occurred with tremendous sound. This is what is our modern "Big-Bang" theory of creation. When Vṛtra died, inertia was broken and the motion starts, i.e. the creation began. So, Vedic speculations are based upon scientific reasoning.

According to water theory of creation, initially there was only water everywhere. Every object of this material world sprang from this water. In the philosophical hymns of the *Atharvaveda*, it is said that the gods originated first from the water. The gods are mostly connected with the element of water.[8] After creation of the gods, the living and non-living elements of this universe were created by them. The modern theory of evolution forwarded by Darwin goes in favour of the water theory of creation given in the Vedas.

Besides the above two main theories of creation, there exist some other considerations about the creation of the material world in the Vedas. The symbolic representation of the nucleus from which the universe was created and which

8. Scherman, *Philosophische Hymnen*, p. 32.

undoubtedly has its origin in the Vedas was adopted by many nations. If we interpret the epithet *haima* or *hiraṇya* (golden) as "bright and reddish yellow" and the words "water" as "water-vapour", i.e. in its gaseous state, we can mark that the story of the golden egg adheres to the notion of a "hot, incandescent nebula" which is the starting point of modern cosmological theories. The epithet *haima* or *hiraṇya* is latent in all the Purāṇas.[9]

Further, references of cosmic dust as the building material of the Universe occur in certain hymns of the *R̥gveda*[10] which are addressed to Viśvakarmā who has designed this Universe by blasting and smelting.

According to well-known Puruṣa-Sūkta of the *R̥gveda*," everything sprang from a giant (*purusa*).[11] In the cosmological speculation of the *R̥gveda*, the sun (*hiraṇyagarbha*) has been regarded as an important agent of creation.[12]

VEDIC THEORY OF GRAVITATION

The laws of gravitation found a place in the earliest theory of the creation of the Universe formulated by Kant, the great Prussian philosopher (1755). According to him, initially there was cold motionless cloud of matter that became, in due course, a vast nebula, spinning so rapidly that rings of matter were successively thrown off its equator by centrifugal action, while the residual central mass remained as the sun. Each ring was supposed to have condensed as a planet, which by a small-scale repetition of the same process threw off satellites, ultimately in this way, the entire solar system was formed. Though Kant's hypothesis is dynamically unsound, still it is

9. S.M. Ali, *The Geography of the Purāṇas*, New Delhi: People's Publishing House, 3rd edn., 1983, p. 192.
10. *R̥gveda*, X.72.2 (Griffith's tr.), p. 496.
11. Ibid., X.90.
12. Ibid., X.121.

important in the sense that the speculation hints at the bottom of the existence of Cosmic nucleus (the Golden Egg) that contained the matter and energy (Heat). Moreover, Kant's hypothesis is the forerunner of the very similar and justly famous nebular hypothesis of Laplace (1796) on which modern speculations on the theory of the creation of the universe rest.[13] This concept of Golden Egg is also latent in the Vedic hymns as described earlier.

The concept of Varuṇa in the *R̥gveda*[14] is essentially the Universal force that has set an order. This cosmic order can only be the law of gravitation.

The cosmic order *r̥ta* governs the entire Universe. According to Murthy,[15] *r̥ta* is the law of the universe or the great cosmic order by which the whole of the manifest universe is working. This, in other words, appears to be the law of gravitation in simple terms. Varuṇa is regarded as the Lord of this law. Many years after, our scientists became successful in recognising the cycle caused due to the law of universal gravitation. The sun strengthens the Universal law, i.e. of gravitation. This is evident from the *R̥gveda*.[16]

VEDIC THEORY OF RELATIVITY

The relative motion of celestial bodies with respect to one another suggests the general theory of relativity as formulated by Albert Einstein.[17] In 1915, Einstein proposed his general theory of relativity in which the framework of the special theory is extended to include gravity, i.e. the mutual attraction of all massive bodies. Whereas the special theory has been

13. S.M. Ali, op. cit., p. 192.
14. *R̥gveda*, V.85.1-6.
15. S.R.N. Murthy, *Vedic View of the Earth*, New Delhi: D.K. Printworld, 1997, p. 81.
16. *R̥gveda*, vi.50.14-15.
17. Fritjof Capra, *The Tao of Physics*, Flamingo, London, p. 72.

confirmed by innumerable experiments, the general theory has not yet been confirmed conclusively. However, it is so far the most accepted consistent and elegant theory of gravity and is widely used in astrophysics and cosmology for the description of the universe at large. Evidently, the paths of celestial bodies have known as geodesics as a consequence of space-time continuum. The mathematical reasoning of the general theory of relativity has been observed by Eddington,[18] who says:

> The Universe is composed of atoms which contain smaller basic particles. These fundamental particles of the atoms are the real essence of the material world, the constituents of the physical Universe. Further, Ronan[19] concludes that both Eddington and Einstein had much in common with Pythagoras and Plato, who also believed that the clue to the Universe is laid in the numbers. The origin of the theory of numbers is due to Vedic sages[20] and it is a result of understanding the laws of the universe through mathematical logic.

EINSTEIN'S MASS-ENERGY RELATION

The philosophy of the Purāṇas revolves around the three main principles of existence, i.e. creation, preservation and dissolution, and which in turn give the notion of the cosmic cycle, i.e. the incidental dissolution of the material world at the end of definite periods (*Kalpa*s of the Purāṇas) and its recreation thereafter. This is quite similar to Einstein's theory of mass transforming to energy and vice versa. The matter can be annihilated to become energy by the famous mass-energy relation of Einstein, which is $E=mc^2$, where m is the

18. C.A. Ronan, *The Astronomers*, London, 1964.
19. Ibid.
20. S.S. Satyaprakash, *Founders of Science in Ancient India*, 2 vols., Delhi, 1990, p. 675.

mass, *E* is the energy equivalent to mass and *C* is the speed of light in vacuum. This mass-energy transformation principle is the outcome of Einstein's Special Theory of Relativity.[21]

GEOMAGNETIC FIELD AND GEOTHERMAL FIELD

Indra of the *Ṛgveda* stands for the natural geophysical phenomenon. Due to interrelationship of Indra and other geological agencies, concepts like *Indrāgni, Indramarutaḥ*, etc. came into existence. The other geological agencies like Agni, Varuṇa, Maruts, etc. are depicted in the Vedas. Indra represents the geomagnetic field which can bring about earthquakes. Agni stands for geothermal field (heat energy), Varuṇa for Universal law of gravitation and Maruts for the wind energy and so on.

UNIFICATION OF FORCES

The Universe is governed by only four forces, viz. gravitational, electromagnetic, weak nuclear force and strong nuclear force. The Vedic bards finally observed that all these forces had equal identity. Then, they observed that all forces are the different aspects of one and the same fundamental force. This ultimate reality is arrived at and is identified with the *Brahman*. Albert Einstein, towards the last part of his life, had tried to unify all these forces. But, he could not succeed in this venture. Later, other scientists of the world tried to unify the four forces. Abdus Salam and other two physicists could succeed in unifying the electromagnetic and weak nuclear interactions. They got Nobel Prize for this.

The divine fire Agni symbolises the geothermal field. At least 200 hymns of the *Ṛgveda* eulogise this deity Agni. Agni has a triple origin — on earth, in waters and in the sky. Agni

---

21. K.T. Achyuta, *Vanamālā Commentary on the Taittirīya Upaniṣad Bhāṣya*, Srirangam, 1913, p. 267.

gives form to all things in nature.[22] It is the energy that creates matter and life. It is the energy in the embryo which forms life-giving cells. Once a life is created, it is that energy which gives the sensation of sight, hearing and thermal condition of a material body. Agni is the supernatural power that exists in everything, either living or non-living. It also creates mental vibrations in every living organism, which, in turn, enable it to perceive the happenings of self and surroundings.

SPACE-TIME CONTINUUM

The word *antarikṣa* in the Vedas means the space extending outward from the earth. The Vedic bards considered the duo of the earth and sky when they spoke of Dyāvāpṛthivī in many Sūktas.The earth and sky are connected by space having air and atmosphere. According to the *Taittirīya Upaniṣad*, it is characterised by the quality of sound. It is noted to give rise to the earth through *vāyu*, *agni* and *āpas*.

The word *kāla*[23] in the Vedic literature generally means time. The general division of time like *bhūta* (past), *bhavat* (present) and *bhaviṣyat* (future) exists in the Vedas. Other divisions of time like *āhna*, *māsa*, *pakṣa*, *saṁvatsara* or *parivatsara*, *muhūrtta* and a larger time interval *yuga* are used in the Vedic literature.[24]

The concept of space and time is significant in Astronomy and Astrology. There is relative importance of space and time in Indian philosophy. Space and time are taken as separate entities by the Nyāya-Vaiśeṣika School. Both are eternal and all-pervasive. They are co-existing with the entire matter, all events and actions. Space, time and matter all have sprung from one cosmic and primordial source called *prakṛti*. The other view predicts two levels of space and time, viz. *akhaṇḍa* and

22. S.R.N. Murthy, op. cit., p. 99.

23. *Ṛgveda*, X.42.9.

24. S. Biswal and B.L. Ray, *Vedic Science and Tecnology*, New Delhi: D.K. Printworld, 2009, pp. 30-31.

sakhaṇḍa. The *akhaṇḍa* is one continuous block of time and space, while the *sakhaṇḍa* is the divisible of space and time. The first represents the nature of *prakṛti* and the second symbolises *ākāśa* under certain conditions. As illustrated by Murthy, each moment is all-pervasive but not eternal. Every moment perishes to room for the next moment. Time is envisaged as a close associate of continuous evolution of matter and is known intuitively.[25] The reflection of *Brahman* is the space.

Modern science defines space as a medium for light to propagate and that bend near celestial objects. The concept of space as gleaned from the theory of relativity of Einstein, it is four-dimensional (x, y, z, t) where time has been taken as one of the dimensions, i.e. space is dependent on time. Hencc, thc space-time continuum is the most fundamental concept that holds the modern scientific thoughts and explanations. This is the object that generates and destroys the material or visible Universe.

VEDIC WAY OF CALCULATING THE SPEED OF LIGHT

In order to calculate the speed of light by the units of distance and time used by the Vedic people, we will follow Sāyaṇācārya's interpretations of the Vedic hymns. Raghava Rao says that Sāyaṇa gives us concrete idea about the speed of light in ½-*nimeṣa*. Sunlight travels 2202 *yojana*s in ½ *nimeṣa*. One *nimeṣa* is equal to 16/75th of second. Hence, the speed of light can be calculated as follows:

In ½-*nimeṣa*, i.e. 8/75 second, light travels 2202 *yojana*s.

1 *yojana*=4 *kosa*s

1 *kosa*=2000 *daṇḍa*s or fathoms

1 *daṇḍa*=2 yards.

1 *yojana* =2×2000×4 yards

=16000/1760 miles (1 mile = 1760 yards)

25. S.R.N. Murthy, op. cit., p. 100.

=100/11 miles.

2202 *yojanas* = (100/11)×2202 miles.

Thus, in 8/75 seconds, light travels 220200/11 miles

In 1 second, light travels (220200/11)×(75/8) miles = 1, 87, 670 miles.

In this manner, the speed of light comes as 1,87,670 miles/second, the modern value being 1,86,000 miles/second. In fact, in CE 1887 Michelson and Morley conducted experiments to determine the absolute speed of the earth through the hypothetical "ether" medium which helped in determining the speed of light.

ASTRONOMY

During the period of the Vedas, astronomy was in a very elementary stage. The naked-eye astronomers of this time meticulously observed the movements of the sun, the stars and the moon. Weber points out that the observations of the sky was only confined to a few fixed stars, especially to 27 or 28 constellations (*nakṣatra*s) and to various phases of the moon.[26] The Vedic concept of *nakṣatra*s also expresses a lunar Zodiac or a system of lunar mansions.[27] A complete set of 28 lunar mansions is noticed in the later *Atharvaveda*.[28]

The Vedic deity Savitā stands for the Sun which effects the day and night due to its movements in the sky. In the solar system, 6 planets are mentioned. These are Mercury, Mars, Venus, Earth, Jupiter and Saturn. The periods of revolution of these planets are also mentioned. The earth (*pṛthivī*) is regarded as circular, being compared with a wheel. Macdonell marks that the heaven (*div*) also stands for *vyoma*

26. A. Weber, 1878, *The History of Indian Literature*, London, p. 246.
27. Ibid., p. 248.
28. *Atharvaveda*, XIX.7.

(sky) or as pervaded with light, the "luminous space". The term *dyāvāpṛthivī* couples both the heaven and the earth.[29]

3. The Atmosphere in the Veda

The air or intermediate space (*antarikṣa*) is described as the region of mists and clouds and it is also called *rajas* meaning watery.[30] Three divisions of the atmosphere have been mentioned. These are *uttama, parama* and *tṛtīya*.[31] The two lower divisions are within the range of our perception, but the third belongs to Viṣṇu.[32] The atmosphere is often described as the abode of celestial waters.

Conclusion

Thus, the physics depicted in the Vedas has relevance with our present-day science with some crude notions of the Vedic seers. From Big-Bang to Big-Vision, the Vedic literature is the storehouse of such concepts and speculations which explain scientific terms developed long after. Starting from Newton's law of gravitation to Einstein's mass-energy relation, the Vedas present pre-scientific knowledge, which, later on, is defined as "Science". The magico-religious phenomena described in the Vedic literature gave rise to the growth of ancient Indian science, which forms the root of all developments in science in true sense.

---

29. *Atharvaveda*, II.27.
30. Ibid., I124.
31. Ibid., IX.22, III.30, IX.74.
32. *Atharvaveda*, VII.99.

6

Concept of Agni in the Ṛgveda in the Light of the Physical Sciences*

Soma Basu and *R.K. Saha*

In the first part of the paper an attempt has been to realise the fact that it is through a proper understanding and realisation of history that the inherent branches of science reveal themselves and in the second, how scientific was the vision of the people of ancient past who not only looked into life and the world with the eye of a poet-artist but also brought to light modern and advanced ideas through simple observation expressed in simple language that wondered the world in those days of hoary past.

Introduction

IF we would like to talk about scientific elements as embedded in the *mantras* of the *Ṛgveda* and that is exactly what has been attempted in this paper we would better keep the following warning of Acharya Brajendranath Seal in mind and proceed to explain the most ancient text in the light of sciences without distorting its true purport or essence by our own made-up ideas, rather we should try our best to bring out the elements of science in whatever rudimentary form they may be found by keeping intact its true spirit.

This is the warning of Acharya Seal who once said,

---

* This is actually a part of the UGC's major research project of the School of Vedic Studies of Rabindra Bharati University, Kolkata.

> I have sought to guard myself against . . . the unscientific, unhistorical but very common and almost inevitable habit of reading modern ideas into old guesses or speculations of a happy-go-lucky or nebulous character. I do not think that the macro passion for Truth is a sufficient safeguard against this fatal facility of unconscious distortion or misrepresentation. A true historical perspective can only be acquired by historico-comparative studies with the application of the correct historico-comparative method.

There is none denying the observation and analysis of the hardcore scientists that even a scholar belonging to the Humanities group of studies could help a lot in tracing ancient India's achievement in science by taking up his or her research in this direction. Reason is or should really be the basis of knowledge. The ancient world (in which I place India in the foremost place) held religion to be the supreme guiding force of man's life here and hereafter. Science progresses by observation, inference, experiment, and correction by falsification towards truth. Humanities also do so. Investigation precedes all beliefs, theories and inventions. The same methods of enquiry were and should even be adopted in philosophy. Ancient philosophers had a philosophical attitude and scientists too had the same kind of outlook. Socrates mastered all the philosophy of the time, which consisted largely of various attempts to provide a scientific explanation of the origin and structure of the universe in contrast to the mythological accounts, accepted by the people. Aristotle, primarily a philosopher, did considerable scientific work particularly in zoology. Knowledge is all-encompassing and frequently the artificial boundary imposed by scholars break. The different spheres merge and again diverge. Let the words "to follow knowledge like a sinking star" of Tennyson be our motto once again.

While emphasising on the contribution of India to the world of knowledge, we particularly feel proud to show that in some fields like the different branches of mathematics, geometry, trigonometry, astronomy, ecology and many more, it is India which contributed the earliest concepts which subsequently in much later ages came to be discovered or brought to light in other parts of the world. Architecture made some advance in Ṛgvedic India. There are references to spacious palace supported by a thousand columns and provided with a thousand doors, e.g. *ṚV*, 2.41.5 says *dhruve sadasyuttame sahasrasthūṇa āsāte* — in supremest home, the thousand-pillared, firmly based and also *ṚV*, 5.62.6 says *sahasrasthūṇa*. Mention is also made of stone castles and structures with a hundred walls, e.g. *ṚV* 10.99.3 says *śatadurasya* which means *śatadvārasya*, i.e. with the hundred gates. Engineering in the Vedic age should be viewed in terms of the making of chariots, construction of houses, pottery, smith, carpentry, etc. The medical art of the age distinguished quite a number of diseases. The use of iron legs as a substitute for natural ones points, however, to some advance in surgery. The science of astronomy made definite progress, and certain stars had already been observed and named.

Many of us know about J.D. Bernal's *Science in History* (when published in 1969 gave a new dimension of approaching history in the light of science), Martin Bernal's *Black Athena* and Rachel Carson's *Silent Spring* (since its publication in 1962 shook the world by verbalising the idea of the fatal effect of pesticides on the animal kingdom) but how many of us care to know the subject matter as contained in the encyclopedic work called *Brāhmasphuṭasiddhānta* (CE 628) of Brahmagupta (CE 598)? In as ancient a period as 1000 BCE a philosopher called Caraka could enumerate the presence of the mind, i.e. *antar indriya — atindriyaḥ punar manaḥ*. He declared that all the sensory organs work in perfect harmony and it is the mind — the supreme organ — which assimilates all the data provided by different sense organs and connects them with the brain.

In the Vaiśeṣika School of philosophy, it has been stated that the mixture of two ever-contradicted substances emits such energy, which is most powerful. The mixture of water and fire, as we all know, emits steam and the power of steam I need not discuss. Interestingly, this is also discovered and documented much earlier than James Watt, who incidentally discovered the steam engine by implementing the power of steam.

The phenomenal growth of modern scientific culture began with the process of deriving knowledge of the past — past is to be accessed with a forward view which is a scientific method. Scientific tradition should be the method of looking into the past — a method, which certainly involves scientists, historians and philosophers. History of science was written in Europe not by a scientist but by a historian and a philosopher. Modern history and philosophy cannot be studied without the history of science. According to Western philosophers like Bacon, Kant, Descartes, it is by following the method that science can be studied and then only appears knowledge. Much mathematical precision is contained in the ancient Indian knowledge, the texts of many of which are written in Sanskrit. They were first translated into Arabic and Persian and then in Latin and then only the European could know about it.

In the context of *R̥gveda* Prof. Winternitz says —

> . . . the songs of the *R̥gveda* prove indisputably that the most prominent figures of mythology have preceded from personifications of the most striking natural phenomena.
>
> — *History of Indian Literature*, vol. 1, p. 77

In the hymns of the *R̥gveda* as presented by the seers no deity is purely divine or purely phenomenal in its manifestation. In all the hymns, the divine and the phenomenal aspects of deities are described together by the seers. The divine was conceived by man through its revelation in nature. The phenomenal or physical aspect of the deity is not divine but represents only the character of the deity. To understand the hymns of the *Ṛgveda* the interpretative concept of *ādhidaivika, ādhibhautika* and *ādhyātmika* is very important and significant:

(a) *ādhidaivika* — the realm of the deities who, as the primary and universal self-manifestations of the divine, are the inner principles or power of nature

(b) *ādhibhautika* — the realm of physical nature in all its aspects, and

(c) *ādhyātmika* — the inner reality of man.

Since the days of the Brāhmaṇas, efforts are still continuing to unearth the idea and for proper interpretation of Vedic Saṁhitās which is one of the oldest and beautiful parts of the corpus of Vedic literature created by mankind. Leaving aside the divine aspect we have put our attention to *ādhibhautika*, i.e. the phenomenon of physical nature of Vedic gods as embedded in the *Ṛgveda* and to see the scientific idea in the vision of Vedic seers.

Agni, one of the principal deities of the *Ṛgveda,* is the divinity which has kept itself continuously important throughout the history of Brahmanism.

Physical (Non-Divine) Attributes of Agni

ALL THAT IS WARM IS AGNI

According to Vedic hymns, Agni manifests itself in natural phenomena as fire, i.e. heat or light. It is seen in Vedic literature that all that is warm is Agni or wherever and in whatever form the principle of heat is present that is all Agni. In many *mantra*s of the *Ṛgveda*[1] Agni is addressed as well as described as *tapurjambhaḥ, tapuḥ, tapiṣṭhaḥ, tigmahete, tigmajambhas, tapiṣṭhena śociṣā, tigmaśocis* and many more, which Sāyaṇa

1. *ṚV*, 1. 36.16 ; 1.58.5 ; 2.4.6 ; 4.4.1, 4 ; 4.5.4 ; 4.15.5, etc.

interpreted as *tapyamanaraśmiyuktāgne, tapiṣṭhaḥ taptṛtamaiḥ tejobhiḥ*, etc. and Wilson translated them as "with burning rays" etc. (Wilson on *ṚV*, 1.36.16). In the *Śatapatha Brāhmaṇa*[2] while describing Agni, it has been said that *yac chuṣkam tad āgneyam* which is translated by Eggeling as — what is dry that relates to Agni.[3]

Seven-rayed Light

In about 65 *mantras* visualised by about 25 seers Agni has been praised as having different colours (rays) with a maximum of seven different colours (rays). In *Ṛgveda* (1.146.1) Agni's epithet *saptaraśmim* means "having/with seven rays" — Sāyaṇa's interpretation is *saptajvālam*; Wilson's translation is "seven-rayed". Again . . . *ā yasmin saptaraśmayas tatā yajñasya netari* (*ṚV*, 2.5.2) expresses Agni as *saptasaṁkhyākā raśmivad* (Sāyaṇa), i.e. "in whom seven rays are manifested".

Agni is visualised as

tamīṁ hotāram ānuṣak cikitvāṁsaṁ ni ṣedire |
raṇvaṁ pāvakaśociṣam yajiṣṭhaṁ sapta dhāmabhiḥ | |
— *ṚV*, 4.7.5

in which the expression *sapta dhāmabhiḥ* is interesting so far it has been interpreted by Sāyaṇa as *saptabhis tejobhir yuktam* and translated by Wilson as "with seven flames" which means of seven different wavelengths, i.e. of seven different energies. Interestingly enough in another *mantra*[4] the expression *dhāma* was commented upon by Sāyaṇa as *dhāmāni tejāṁsi* which Wilson has translated as "resplendent flames".[5]

2. *ŚB*, 1.6.3.23.
3. P. 169, tr. J. Eggeling (Sacred Books of the East Series).
4. *Ṛgveda*, 6.2.9.
5. We would like to add here as to why the expression *dhāman* has been taken in the sense of "energy" though the primary meaning of the word being "abode" (vide Monier-Williams' Dictionary). In it as well as in the *Vācaspatyam* the secondary meaning of *dhāman* →

Various epithets of Agni as *rohitā*[6] and *rohitāsa,*[7] *harikeśa?*[8] and *harivrataṁ,*[9] *śītipṛṣṭha,*[10] *nīlapṛṣṭha,*[11] *āruṇīḥ,*[12] *hiraṇyarūpam,*[13] etc. meaning "red-coloured", "tawny-rayed" and "tawny-hued", "black-backed", "purple dawn" and "golden-formed" show that to Vedic seers Agni, i.e. light (visible) was not monochromatic, i.e. of single colour or ray but Agni, i.e. light is chromatic, i.e. composed of different colours or rays. We know from modern-day observations and measurements that a luminous object like the sun, star or a luminous body emits radiations and the rays of the radiation spectrum depend upon the temperature of the luminous body. To be pragmatic we can safely say no sophisticated instruments were available in those days of hoary past to measure or have any idea of left and right region of the visible spectra. In the Vedic ages the seers were aware of only the visible portion of the total spectrum of electromagnetic radiation from luminous Agni and we know the range of wavelength of visible spectra is from 4000°A to 8000°A and there are seven different colours in the visible region of the spectrum. Newton from a series of experiments concluded that the sunlight is composed of a number of coloured lights and this polychromatic light consists of seven colours, i.e. more than one colour. By allowing a

→ has been cited as "energy" and we have taken the word as meaning "energy" as the suffix attached is *bhis,* i.e. *dhāmabhis,* i.e. instrumental plural which gives the sense "with" and which goes perfectly with the holistic meaning of the *mantra*. If we take the word *dhāman* as meaning "abode" then it would have taken locative, i.e. 7th case ending and the word in that case would have been *saptadhāmasu.*

6. *Ṛgveda,* 1.94.10.
7. Ibid., 4.6.9.
8. Ibid., 3.2.13.
9. Ibid., 3.3.5.
10. Ibid., 3.7.1.
11. Ibid., 3.7.3.
12. Ibid., 4.1.16; 2.16.
13. Ibid., 4.3.1.

narrow beam of white light from a luminous object pass through a prism he showed that the refracted beam striking a white screen produces on it an elongated coloured band of light having seven different colours: violet, indigo, blue, green, yellow, orange and red proving that the ray is actually composed of seven different rays having seven different wavelengths, i.e. seven different vibrations, i.e. seven different energies. Newton's experiment of seventeenth century CE proved unambiguously the veracity of meticulous observation and imagination of Vedic seers.

Agni as the Supporter of the Universe (*R̥V*, 5.7.6)

Some seers visualised Agni as the "supporter of the universe," oft-quoted epithets being *dhartā, vidhartā, viśvasya dhārakaṁ;* Sāyaṇa's interpretation is *sarvasya jagataḥ dhartā* and Wilson's translation is "supporter of the universe"; *yaṁ martyaḥ puruspr̥haṁ vidad viśvasya dhāyase* that means — man recognises Agni as the sustainer of all and Griffith translates as — whom . . . mortal man hath found to be the stay of all — *sarvasya vyāpārārthaṁ.*

From all these we can conclude that the concept of Agni, i.e. heat energy imagined and given expression by the Āryans for sustaining the universe and of its presence everywhere is not only a bold one revealing one of the most important scientific truths on presence of heat energy everywhere, but proves their minute observational skill of ending up on a chapter on kinetic theory of heat deduced and proved by the physicist in a very later stage.

It is proved from modern experiments (Rumford and Davy) that heat is a form of energy and due to a kind of motion of particles is directly proportioned to absolute temperature. According to this theory, every substance — solid, liquid or gas — consists of myriads of tiny particles called molecules which are the ultimate unit of substance capable of existing in free-state. These molecules possess all the characteristics of the substance. They are always in motion, their velocity increases with temperature. In 1859 Maxwell deduced the law of equi-partition of energy which states that in a dynamical system in thermal equilibrium the energy is equally distributed between the various degrees of freedom and for each of this it is equal to ½ kT where "k" is the Boltzman constant and T is the equilibrium temperature of the system in Absolute Scale. This was applied successfully by Boltzman for the rotational and vibration system. The above relation clearly shows for sustaining the universe, the energy for each degree of freedom must have an energy value ½ kT and is proportional to heat (temperature). As "k" is constant, the absolute temperature "T" which is a measure of heat must have some positive value otherwise for zero value of energy or heat, i.e. temperature, all will come to a standstill. So sustaining of universe without temperature is absolutely impossible. The question of having negative value of "T" does not arise. The presence of Agni, i.e. heat/temperature everywhere is a must for survival of the universe — a beautiful truth imagined by the Vedic seers.

Agni as Present Everywhere (*ṚV*, 1.70.2)

Ten different seers of about 20 *mantra*s of the *Ṛgveda* have visualised the presence of Agni everywhere, i.e. within water, all moveable and immoveable things, it being the cause of sustaining life in the universe and is responsible for the growth of herbs, shrubs and plants, ripening of grains, presence in the wombs of all living beings and is responsible for the digestion of food (*jāṭharāgni*) on which depends sustenance and growth of the bodies.

> *garbho yo apāṁ garbho vanānāṁ garbhaśca sthātāṁ garbhaś carathām* |
> *adrau cid asmā antarduroṇe viśāṁ na viśvo amṛtaḥ svādhīḥ* | |
> — *ṚV*, 1.70.2

> (They offer oblations) . . . to that Agni, who is within the waters, within woods, and within all moveable and immoveable things. . . .
>
> He is the embryo, the internal germ of heat and life:

As Sāyaṇa interprets it as — *agniḥ apāṁ garbhavad antarvartī araṇyāṁ madhye varttate sthāvarāṇāṁ jaṅgamānāṁ jāṭhararūpeṇa dehamadhye avatiṣṭhate.*

Why Agni was Addressed as "Messenger" or "Oblation-bearer" in About 110 Mantras of the Ṛgveda?

Maximum number of times Agni has been praised as performing the function of messenger or oblation-bearer (about 110 *mantras* by about 34 seers) and sometimes both —*dyūtaṁ*[14] (Sāyaṇa — *devānāṁ dyūtasthānīyaṁ* — *ṚV*, 1.44.11), *vārtāharaṁ, havyavāhana*[15] (Sāyaṇa — *havyānāṁ voḍāraṁ, haviṣo voḍaḥ,* etc.), *agnir dyūto abhavad havyavāhano* (*ṚV*, 5.11.4) (Sāyaṇa — *havyānāṁ haviṣāṁ voḍā agnir devānāṁ dyūtaḥ abhavat*) which means "Agni, the bearer of oblations, has become the messenger."

Agni as Travelling in All Directions (*ṚV*, 1.145.1)

According to at least 17 seers of about 36 *mantras* Agni travels in all directions, i.e. he is *viśvāyuḥ,*[16] "he goes everywhere" — *īrate,*[17] *sa jagāmā,*[18] etc. Sāyaṇa interprets as *vyāptagamanaḥ, sa sarvatra gacchati,* etc.

Agni as Most Rapid (*ṚV*, 4.4.3)

Agni is *ajiraṁ,*[19] *tūrṇitamo,*[20] *urujrayasam,*[21] i.e. "most rapid".

14. *Ṛgveda*, 1.12.1; 1.36.3; 1.44.3; 1.44.11; 1.60.1; 5.6.8, etc.
15. *Ṛgveda*, 1.36.10; 1.44.5; 3.9.6; 3.29.7.
16. Ibid:, 1.27.3.
17. Ibid., 1.140.5.
18. Ibid., 1.145.1.
19. Ibid., 3.9.8.
20. Ibid., 4.4.3.
21. Ibid., 5.8.6.

Sāyaṇa interprets as *kṣipraṁ gantāraṁ, tvaritatamaḥ, prabhūtavegaṁ.*

Agni as All-Pervading (*ṚV*, 6.15.8)

Agni is *vigāhaḥ,*[22] *vibhuḥ,*[23] etc. that is "all-pervading", i.e. *sarvatra vyāptam.*

On analysing these above-mentioned findings we cannot but wonder by the brilliance of the Vedic seers' wisdom in selecting Agni to act as messenger among all deities for sending their message of invitation to other gods in heaven, mid-air and earth and bring them to the sacrificial altar and also for carrying sacrificial oblations to gods in three regions. Now the question is why this selection of Agni as messenger or oblation-bearer while other gods endowed with great physical attributes like the ever-moving (*satatagamanaśīla*) Maruts and the most powerful Indra were left out to do the same? Modern science has shown that nothing can travel faster than light. When the velocity of sound is 3.5×10^4 cm/second, light travels with a velocity of 2.99×10^{10} cm/second. Light takes about 8.25 second to travel from the sun to the earth. Moreover thunder or sound needs a medium for propagation but light does not. It can travel to any region — celestial or on the earth — most rapidly and without any medium. In 1690 Huygen a Dutch Physicist presented a theory and proved that light travels in the form of waves in all directions. As the propagation of light occurs also in vacuum Huygen postulated vehicle for these waves an omnipresent, all-pervading, universal and continuous medium called luminiferous ether. So the selection of Agni, i.e. light as accepted messenger of the gods and bearer of oblation by various seers in the *Ṛgveda* confirms the presence of pragmatic realisation of a scientific truth in their imagination and observation that Agni, i.e. light is the fastest moving, travels in all directions and all-pervading which is imagined and proved by the physicists at a much later stage.

22. *Ṛgveda*, 3.3.5.
23. Ibid., 6.15.8.

So it will not be far from truth if we conclude from our observation and study of the *mantras* of the *Ṛgveda* that the Vedic seers in praising Agni in so many *mantras* of the *Ṛgveda* have spoken of a number of natural laws and physical phenomena of Agni, i.e. fire/heat/light which are found to be in conformity with the idea of modern-day science. Though there is none denying the fact that the Ṛgvedic deities were primarily and basically nothing but personified natural phenomena, yet the prime focus of imagination, thought and vision of the seers was to praise the divinity or divine qualities inherent in these deities. Meticulous observation and experience led the seers to unfurl the philosophy of basic science governing the natural laws in their *mantras* in praising deities and we can see the glimpses of scientific truths embedded in their vision of physical attributes of the deities who are personifications of various natural phenomena.

In conclusion we see our ancient sages from their meticulous observations and imagination could intuitively attain knowledge in certain forms which took a long time for the Europeans to arrive at in the present form of natural science. Thoughts of Vedic seers were scientifically significant which were lost in the midst of antiquity as because proper attention and investigation were not duly emphasised. Praises and prayers expressed by them through the *mantras* of the Vedas acted as entry for them to supernatural through the deities and in their praises we find the principles of fundamental science remaining embedded in whatever rudimentary form. To take it further we must work and give our highest attention in bringing to light and highlighting the importance of our national heritage — a brilliantly intellectual corpus of Vedic literature — a true glory of our past.

This is only a glimpse of what we have found out about Agni interpreted, and tried to correlate and bring to light, a large part still remaining to be placed in front of you.

Appendix

We have also worked in the same line by taking up the Maruts, a glimpse of which I give in this Appendix.

Vedic people had a clear-cut idea about power, the force of terrible strength and its effect and also magnitude of the strength of the wind, the effects of the Maruts (i.e force) and its reaction though they might not have the idea of measuring strength and its unit. The Maruts are powerful, possessor of vigour, strong, violent, the symbol of power with tremendous capacity and great strength for doing work, shakers/agitators of the earth. A force of tremendous high magnitude can "agitate", "shake" and "tremble" any object like mountains, earth and ocean.

In our study we have come across some *mantra*s wherein the concept of like parallel forces — an important part of statics — is embedded. Further the concept of not only parallel forces but also of like parallel forces and bending due to couple is embedded in the physical attributes of the Maruts as found in the *Ṛgveda*. The idea of bending arising due to couple (i.e. two unlike equal parallel forces, one due to the Maruts or wind and opposite one from the ground) can be seen as containing in the description of the physical features or attributes or properties of the Maruts as visualised by the Vedic seers. Vedic seers' addressing of the Maruts always in plural is not only rational but a perfectly correct representation of the phenomenon of forces in such cases. Mathematically the effect of the Maruts, i.e. wind cannot be on a single point. Also we find the idea of resultant force in physical attributes of the Maruts and the most important feature the presence of the concept of origin of sound as visualised by the Vedic seers. It is astonishing to realise that when the seers say — while the Maruts pass along, they talk (the root used is *brū* — 'to speak', 'to talk') 'together on the way', they recognise unconsciously the phenomenon of force (of the Maruts or wind) setting air — medium in vibrations, i.e. the same phenomenon behind the generation of sound in a flute.

Coming to the physical attributes of the Storm Deity, i.e. the Maruts, which is next to Indra amongst atmospheric gods, we would like to analyse and present here some of them as found in the hymns of the *Ṛgveda* and will see these attributes in the eyes of preliminary science. The Maruts are celebrated in the *Ṛgveda* independently in 33 hymns and conjointly with Indra in 7 hymns and 1 each with Agni and Pūṣan. They are always celebrated in plurals (*gaṇaṁ tavasam* — *ṚV*, 5.58.2). The physical attributes of the Maruts, as visualised by the Vedic seers in those *mantra*s related to the Storm Deity, the Maruts, give us an impression of their thoughts and ideas on the Maruts wherein it is seen preliminary ideas of modern-day statics and dynamics and idea of origin of sound are embedded.

In our study we have seen in about 32 *mantra*s related to the Maruts of the *Ṛgveda* more than 12 seers viewed the Maruts as powerful, possessor of vigour, strong, violent and the like, i.e. they depicted the Maruts as the symbol of power. It has tremendous capacity and great strength for doing work. In several *mantra*s the Maruts are termed as *śavasāhimanyavaḥ*[24] (on which Sāyaṇa commented as *balena āhananāśīlamanyuyuktāḥ* and Wilson translated as "of deadly strength"), *taviṣībhir āvṛtaḥ*[25] (Sāyaṇa — *anyeṣām asādhāraṇair balaiḥ pariveṣṭitaḥ* and Griffith translated as "endowed with power and might", the term *taviṣī* arises from the root *tu* meaning "strong"), *satyaśabasaḥ*[26] (Sāyaṇa — *avitathabalāḥ*, Wilson — "possessor of true vigour"), *ojobhir ugrāḥ*[27] (Sāyaṇa — *balair ūdgūrṇāḥ*, Wilson — "strong with varied power", the term arises from the root *uj* or *vaj* meaning powerful), *śikvasaḥ*[28] (Sāyaṇa — *śaktās te*, Griffith — "mighty"), *yuvānaḥ*[29] (Sāyaṇa — *nityataruṇāḥ*, Oldenberg — "youths"), *śardho mārutam* . . .[30] (Sāyaṇa — *mārutaṁ balaṁ mahad ati pravṛddham*, Wilson — "infinite strength of the Maruts"), *śurā ivet*[31] (Sāyaṇa — *śauryopetā yuyutsavaḥ puruṣā iva*, Wilson — "combined with strength"), *gaṇaṁ tavasaṁ*[32] (Sāyaṇa — *mārutaṁ balavantam*, Wilson — "powerful company"), *ṛṣvāḥ*[33] (Sāyaṇa — *mahānto marutaḥ*, Wilson — "mighty Maruts").

Regarding the main root of the Maruts, it has been stated in Yāska's *Nirukta* as appearing to be *mara* which has got connection with the root *mṛ*. The meaning of the root, among many, may be mentioned as "to die", "to crush" or "to shine". To be specific *mara*

24. *Ṛgveda*, 1.64.8; 9.
25. Ibid., 1.64.10; 1.87.4; 187.1; 5.58.2; 8.7.2 — *taviṣīti bala nāma* — Sāyaṇa.
26. Ibid., 1.86.8, 9.
27. Ibid., 1.171.5; 7.56.6, 7; 8.20.3.
28. Ibid., 5.52.16; 54.4.
29. Ibid., 8.20.17, 19.
30. Ibid., 5.52.8.
31. Ibid., 1.85.8.
32. Ibid., 5.58.2.
33. Ibid., 5.52.6.

coming from the root *mṛ* meaning primarily "dying" or "death" seems to go perfectly in tune with "strength" (one of many physical attributes of the deity) of the Maruts as given in many *mantras* attributed to the Maruts in the *Ṛgveda*. Moreover another root *mṛd* with prefixes like *pari* or *ā* also gives the sense as "to crush", "to grind".

From above references and discussions it can be said that Vedic people has a clear-cut idea about power, strength and force and also magnitude of the strength of the wind though they might not have the idea of measuring strength and its unit.

In our investigation in the independent hymns relating to the Maruts and also in the hymns dedicated conjointly in association with other deities we have come across the *mantras* by 12 seers wherein the Vedic seers visualised the effects of the Maruts and its reaction. Following are some of the expressions in the *mantras* in relation to above-mentioned physical attributes of the Maruts along with Sāyaṇa's interpretation and English translation. *dhūtayaḥ*[34] (Sāyaṇa — *sthāvarādīnāṁ kampanakāriṇo marutaḥ, kampanaśīlāḥ, kampakāḥ,* Wilson — "shakers/agitators of the earth"), *vepayanti parvatān*[35] (Sāyaṇa — *meruhimavad ādīn marutaḥ prakarṣeṇa kampayanti,* Wilson — "They (the Maruts) make the mountains tremble", Griffith — "Whole strength have ye . . . earth-shakers, is your might"), *dhūnayo*[36] (Sāyaṇa — *sapatnān kampayanta iva,* Wilson — "like heroes making (adversaries) tremble"), *eṣāṁ bhiyasā bhūmī rejati*[37] (Sāyaṇa — *eṣāṁ marutāṁ bhayena bhū kampate* and Wilson — "the earth trembles with fear"), *dhruvacyutaḥ*[38] (Sāyaṇa — *dhruvānāṁ niścalānāṁ parvatādīnām api cyāvayitāraḥ* and Griffith — "the Maruts make all things to reel"), *rejate . . . pṛthivī*[39] (Sāyaṇa — *marudbhyo bhayāt pṛthivī kampate* and Griffith — "earth shakes in terror of their wars"), *acucyavurdṛḷhāni*[40] (Sāyaṇa — *dṛḍhānyapi parvatāni*

34. *Ṛgveda*, 1.37.6, 1.39.1, 10; 1.87.3; 1.168.2; 5.54.4; 7.58.4; 8.20.16.

35. Ibid., 1.39.5; 1.64.11; 1.167.8; 1.168.4 (Oldenberg observes, "the object of the poet is to show the strength of the Maruts. Vedic poets do not shrink from the conception that the Maruts shake even mountains".

36. Ibid., 1.87.3; 6.66.10; 8.20.14 (*dhūnīnāṁ*); 10.78.3.

37. Ibid., 5.59.2; 1.39.3; 1.64.3; 11; 85.1; 88.3; 2.34.1; 2; 5.54.1; 3; 57.3; 60.2; 6.66.9; 7.56.19; 57.1; 58.1; 8.7.2; 5; 8.20.4; 5.

38. Ibid., 1.64.11.

39. Ibid., 6.66.9.

40. Ibid., 1.168.4.

cyāvayanti cālayanti and Wilson — "the Maruts . . . have cast down even firm things)".

From these above-mentioned expressions used in the *mantra*s related to the Maruts it is clear that the Vedic seers had a clear concept of the force of terrible strength of the Maruts and its effect. A force of tremendous high magnitude can "agitate", "shake" and "tremble" any object like mountains, earth and ocean.

In about 35 *mantra*s related to the Maruts visualised by nine seers we have seen the concept of not only parallel forces but also of like parallel forces and bending due to couple is embedded in the physical attributes of the Maruts in the *Ṛgveda*.

Vedic seers' addressing of the Maruts in plural is not only rational but a perfectly correct representation of the phenomenon of forces in such cases. Mathematically the effect of the Maruts, i.e. wind cannot be on a single point. The following expressions or portions of the *mantra*s related to the Maruts supports the above idea. The expressions *gaṇam mārutam*[41] (Sāyaṇa — *marutāṁ sambandhinaṁ gaṇaṁ gaṇāvayavabhūta marutāṁ* and Wilson — "company of Maruts"), *ayāsāṁ marutām anīkam*[42] (Sāyaṇa — *gantṛṇāṁ marutāṁ samuhaṁ* and Griffith — "army of the restless Maruts"), *śardho maruto*[43] (Sāyaṇa — *bhavatāṁ gaṇam* and Wilson — "company of Maruts") are some of the like expressions found in the *mantra*s related to the Maruts in the *Ṛgveda*.

In our study we come across some *mantra*s wherein the concept of like parallel forces — an important part of statics is embedded. The expressions *abhi . . . mitho vapanta*[44] (Sāyaṇa — *marutaḥ parasparaṁ saṁgachhante* and Wilson — "they go together by their own pure paths"), *ṛtena satyam ṛtasāpa*[45] (Sāyaṇa — not clear and Wilson — "in proper order came those who truly follow the order"), *gaṇaśriyaḥ*[46] (Sāyaṇa — *saptagaṇarūpenāvasthitāḥ* and Wilson — "Maruts are distinguished in troops"), *vayo na ye śreṇī*[47] (Sāyaṇa — *pakṣiṇa iva paṅktayaḥ*) support our view. In addition to this the idea

41. *Ṛgveda*, 5.52.14; 53.10.
42. Ibid., 1.168.9.
43. Ibid., 5.54.6.
44. Ibid., 7.56.3.
45. Ibid., 7.56.12.
46. Ibid., 1.64.9.
47. Ibid., 5.59.7.

of equal like-parallel forces is found in physical attributes of the Maruts as visualised by the seers. The portion of the *mantra*[48] — *te ajyeṣṭhā akaniṣṭhāsa . . . amadhyamāso* (Sāyaṇa — *marutaḥ ajyeṣṭhā akaniṣṭhāḥ . . . amadhyamāḥ sarvaprakāraiḥ samāḥ* and Wilson — "none of them are older, none younger (than the others) . . . none hold a middle (rank) but all excel in glory"), *ajyeṣṭhāso akaniṣṭhāsa ete saṁ bhrātaro vāvṛdhuḥ*[49] (Sāyaṇa — *parasparaṁ jyeṣṭhakaniṣṭhabhāvarahitāḥ sahaivotpannāḥ samānabalāḥ parasparaṁ bhrātṛbhūtāḥ santaḥ . . . vardhante* and Wilson — "they are brothers, of whom no one is the elder, no one the younger, but . . . grew up together for their mutual prosperity") indicates the concept of equal like-parallel forces in describing the physical attributes of the Maruts.

From *yujā* (*ṚV*, 1.39.4) (Sāyaṇa — *yogena parasparaikamatyena* and Wilson — "May your collective strength be quickly exerted through the union of you"), we find the existence of the idea of resultant force in physical attributes of the Maruts as visualised by the Vedic seers.

Further from the expression *ni yad yāmāya vo girir ni . . . yemire* which Sāyaṇa interpreted as *he marutaḥ yuṣmākaṁ gamanāya girayaḥ parvatāḥ svayam eva niyamyate* (*ṚV*, 8.7.5) and Oldenberg translated as — "mountains bent down before your march. The mountains are curved, the rivers are restrained at your coming, for the upholding of your great strength" and from another expression like *sthirā cin namayiṣṇavaḥ* (*ṚV* 8.20.1) which as per Sāyaṇa's comment is *sthirāṇi dṛḍhānyapi parvatādīni namanaśīlāḥ kampayitāraḥ* and Oldenberg's translation is — "who can bend even what is firm" — we see that the idea of bending arising due to couple (i.e. two unlike equal parallel forces; one due to the Maruts or wind and opposite one from the ground) can be seen as containing in the description of the physical features or attributes or properties of the Maruts as visualised by the Vedic seers.

In our investigations related to the Maruts in the *Ṛgveda*, we have seen presence of the concept of origin of sound in about 19 *mantras* visualised by seven seers. Now it is known and can be demonstrated as well as established easily that a body or medium generates sound only when it is set into vibrations. A good number of *mantras* or parts of *mantras* clearly express the seers' knowledge about the production/generation of sound due to movement of the

48. *Ṛgvedaw*, 5.59.6.
49. Ibid., 5.60.5.

Maruts through air. From their keen observation they could observe the phenomenon that when the Maruts travelled through layers of air, they act as generators of speech, i.e. sound.

It is astonishing to realise that the *mantras* like *yaddha yānti marutaḥ saṁ ha brūvate adhvan nā śṛṇoti kaścid eṣām* (ṚV, 1.37.13) to mean — *yadā khalu marutaḥ gacchanti tadānīṁ mārge sarvataḥ dhvanim avaśyaṁ kurvanti* — that means — the while the Maruts pass along, they talk (the root used is *brū* — to speak, to talk) together on the way (as translated by Griffith) or *yad vadān* (ṚV, 1.37.3 the root again is *vad* — to speak, to talk) or *vācam udīrayanti* (ṚV, 1.168.8) the root again is *ut-īr* — to speak, to talk), indicate directly but unconsciously the phenomenon of force (of the Maruts or wind) setting air- medium in vibrations, i.e. the same phenomenon behind the generation of sound in a flute. Here are some of the expressions from the *Ṛgveda*, which show the idea of origin of sound as containing in the imagination of the Vedic seers. *te sūnavo giraḥ* (ṚV, 1.37.10) means the Maruts are "the sons, the singers" according to Griffith and the "generators of speech" according to Wilson. Sāyaṇa explains the phenomenon as *vāca utpādakā marutaḥ, vāyavo hi tālvoṣṭhādiṣu saṁcaranto vācam utpādayanti,* (i.e. the producers of sound). Again the description of the Maruts as *gāyad gātham* (ṚV, 1.167.6) — *gātavyaṁ stotraṁ gāyati* as per Sāyaṇa and — "O Maruts, had sung his song" as per Oldenberg or *vāśrāsaḥ* (ṚV, 8.7.3) *vāśrāsaḥ* (ṚV, 8.7.7) or *vāśati* (ṚV, 5.54.2) or *stanayadmā* (ṚV, 5.54.3) — *śabdakāriṇaḥ* or *śabdayati* or *śabdopeteganaḥ* meaning "loud-roaring or loud-sounding winds" or "roaring in concert" — as translated by Griffith and Wilson. Instances can be multiplied where this very attribute of the Maruts as making sound has been repeatedly spoken of. In the expressions like *virapśinaḥ* (ṚV, 1.64.10; 1.87.1; 1.166.8) the root being *rap* or *lap-vyakta śabda* and the derivation given by Sāyaṇa himself being *vividham śabdaṁ rapantīti virapśaḥ . . . maruto hi vividhaṁ śabdaṁ kurvante* or *nānadati* (ṚV, 1.64.8) the root being *nad* to mean *avyakta śabda* and interpreted by Sāyaṇa as — *maruto . . . bhṛśaṁ śabdaṁ kurvanti . . . marutsu api āgateṣu gambhiraḥ śabda utpadyate.*

From the above discussions it can be concluded that the idea of force and its effect was embedded in the visualisation of the Vedic seers. This preliminary concept of force was not developed further but was lost in antiquity.

References

Ṛgveda-Saṁhitā (with the commentary of Sāyaṇācārya, Vaidika Saṁśodhana Maṇḍala) Poona: Vedic Research Institute, 1983 (2nd edn.) vols. 1-4.

Ṛgveda Saṁhitā, tr. H.H. Wilson, New Delhi 1977, vols. 1-7.

The Hymns of the Ṛgveda, tr. R.T.H. Griffith, Varanasi, 1971, 2 vols.

Śatapatha Brāhmaṇa, tr. J. Eggeling, Sacred Books of the East Series, 1882, Delhi, 1978.

7

Mechanics in Hindu Literature

P. Priyadarshi

Although we do not have much surviving science texts from ancient India, the ones which are available show that there was a proper physics taught in India at least since sixth century BCE. This tradition of science ended in the twelfth century. Kaṇāda (sixth century BCE), the great master of the Vaiśeṣika School of Indian philosophy wrote the *Vaiśeṣika-Sūtras,* which contains systematic exposition to physics, probably meant for students. In this article, some of his *Sūtras* have been examined whether they meet modern standards of physics. Only a very limited section of the text was examined. From that, it becomes obvious that Kaṇāda had knowledge of velocity, momentum, gravitation, acceleration, energy, etc. The text anticipates Newton's Laws of Motion. It discusses active and passive work, potential energy, collision, impulse and inertia. There is explicit mention of magnet and magnetic compass in Kaṇāda's text, which other authors have so far thought came to India only in the sixteenth century. Praśastapāda was another scientist of this school who lived in sixth century CE. Some of his texts have also been examined.

Introduction

OWING to large-scale destruction of libraries and academic institutions by invading armies in the twelfth century, most of Indian science literature was lost forever. Yet we get some glimpse of ancient Indian science from many religious texts which survived because they were in private possession of priest class. These texts were not intended to be the books of science. Yet, often the scientific information contained in these texts is substantial. However some pure science texts have also survived and have been unearthed within last two centuries. These provide us the information as to what was the shape of science in ancient India.

The ancient Indians were only just behind the equation formulation stage in mechanics when the Muslim invasion took over India. Probably, had India not gone under the rule of foreign civilisations, most of the modern concepts of mechanics would have been formulated in India much before they were formulated in the West. Bhāskara (Bhāskarācārya II; *c.* CE 1150) was the last great mathematician-physicist of India. He was also the greatest scientist in the world at his time. Bhāskara refined physics, especially mechanics, to the standards of mathematical accuracy. His achievements were incredible. For example, he calculated that the centre of earth is not the centre of orbit of moon's rotation.[1]

Bhāskara gave the concept of *tātkālika gati* (instantaneous velocity). He gave the formula $v = s/t$, where v is the average velocity (*sthūlagati*), s is the distance traversed and t is the time taken.[2] Where velocity is uniform, interval of time may be of any amount (*sthūlakāla*), but when the velocity is variable, an infinitesimally small amount should be taken (*sūkṣmakāla*), i.e. ds/st. It is in this way that Bhāskara calculated the instantaneous velocities (*tātkālika gati*) and instantaneous positions of many planets. This was the birth of differential calculus, the backbone of all future physics, 600 years before Newton.[3] Bhāskara's death in 1185 was the end of a very long tradition of Indian physicists. Muhammad Ghori invaded and captured India in 1192, and destroyed all educational

1. Bhāskarācārya, *Chedyakadhikāraḥ*, 10-21, in *Golādhyāya* of *Siddhānta-Śiromaṇi*.
2. Bhāskarācārya, *Gaṇitādhyāya, Siddhānta-Śiromaṇi.*
3. B.N. Seal, 1958, *The Positive Sciences of the ancient Hindus*, New Delhi: Motilal Banarsidass, pp. 149-50.

institutions, big or small. Credit of burning the famous Nalanda University in 1193 goes to his general Bakhtiyar Khilji.

We do not know when actual systematic study of mechanics started in India. But it must have been much before sixth century BCE. The earliest text available to us which can be labelled a proper physics textbook is the *Vaiśeṣika-Sūtras* by Kaṇāda. Kaṇāda lived in the sixth century BCE, and was the most famous among the Vaiśeṣikas known to the posterity. He has mostly been known to us for his "atomic theory". He was a thorough physicist. His physics is based on "application of logic" to observations. Vaiśeṣikas perfected logic by eliminating logical fallacies and superstitions. In that system there was no place for beliefs. They were staunch "realists". They were thus able to discover many of the truths of science just by means of logic and observation.

In this article we will examine some of the ideas in mechanics from Kaṇāda's *Vaiśeṣika-Sūtras* and Praśastapāda's commentary on that, both of which texts are fortunately extant today.

Kaṇāda's Vaiśeṣika-Sūtras

Kaṇāda discussed properties of matter, and identified that motion was an inherent property of substance (*kriyavattva*).[4] Mass (*parimāṇa*) was another property of matter.[5] In Vaiśeṣika vocabulary, energy has been called *karma*. The Hindu metaphysical concept of *karma* or "accumulated past work" is actually a literal application of a fact of physics, born out of scientific understanding of relationship between work and energy. At least two thousand years before the West, they had the understanding that work done against some force

4. *Vaiśeṣika-Sūtras*, 1.1.15 and 2.1.12; also see S.D. Chatterjee, "Physics and Mechanics in Ancient and Medieval India", op. cit. below, p. 102.

5. *Vaiśeṣika-Sūtras*, 1.1.5.

becomes stored as energy, which can be later released for work. Thus, in approach, the *Vaiśeṣika-Sūtras* of Kaṇāda are quite modern. In the following paragraphs, we shall examine some of the *Vaiśeṣika-Sūtras*. They are being given with meaning and my commentary. Original Sanskrit has been kept intact except that *sandhi*s have been broken (*anvaya*) to clarify the meaning.

dravyāṇi-dravya-antaram-ārambhante guṇaḥ-ca-guṇa-antaram |

One substance changes into another substance and its properties change into other properties

— *Vaiśeṣika-Sūtras*, 1.1.10

[Comments: This *sūtra* has to be read in light of the most basic proposition of Vedic (orthodox) philosophy — the law of conservation of matter and energy or the First Law of Thermodynamics. In other words, matter (and energy) can neither be created nor destroyed. It can only change from one form to another. Form and other manifest characteristics of substance are changeable, but not the basic substance measured today as mass. The law was known to the Vedic people, and we find it in the *Bhagavad-Gītā* in the following verse (*nāsato vidyate bhāvo, nābhāvo vidyate sataḥ* — "That which did not exist, cannot have existence; that which exists can never be non-existent").[6] *Chāndogya Upaniṣad* states, *kathamasataḥ sajjāyet* (how can something existing arise from something non-existing?)[7] According to Bal Gangadhar Tilak, this law and the law of causality were the two fundamentals on which Hindu thought, especially the Sāṁkhya philosophy had been based.[8]

---

6. The *Bhagavad-Gītā*, 2.16.

   "*Na* (no) *asato* (non-existing) *vidyate* (happens to be) *bhavaḥ* (presence, existing)

   Na (no) *abhāvo* (absence) *vidyate* (happens to) *sataḥ* (existing)".

7. *Chāndogya Upaniṣad*, 6.2.2.

8. B.G. Tilak, 2002, *Śrīmad-Bhagavad-Gītā-Rahasya* (Hindi tr. by Sapre, Madhavraoji), Poona: DJ Tilak, 18th edn., pp. 100-101.

Al-Biruni who visited India in *c.* CE 1000 gives a summary of the Hindu ideas of the law of conservation of matter and energy. Al-Biruni mentions that the Hindus believed that matter and energy could neither be created nor destroyed.[9] Al-Biruni writes, "(the Hindus) believe matter (and energy) to be eternal. Therefore they do not by the word "creation" understand a formation of something out of nothing. They mean by creation only the working with a piece of clay, working out various figures and combinations with it."[10] He further adds, by creation Hindus imply only a change of form or features and by destruction also they imply loss of certain features. "By such a creation, not one piece of clay comes into existence which did not exist before, and by such destruction not one piece of clay which exists ceases to exist."[11]

Al-Biruni further writes, "It is quite impossible that the Hindus should have a notion of a creation as long as they believe that matter existed from all eternity."[12] Creation is just "making such arrangements with it (existing matter) as will lead to certain ends and aims which are potentially in it".[13] He adds as far as this is the meaning and limitation of creation, even the human beings are, at least theoretically, capable of creating anything — even living beings — in the Hindu concept.[14] (In Hindu belief) every creation implies change of form of something existing and hence creation necessarily implies destruction of the earlier form of the matter. Hence destruction takes place as kinds of new formation.[15] (The Hindus believe) at the end of the world,

---

9. Q. Ahmad, 1999, *India by Al-Biruni*, New Delhi: National Book Trust, pp. 146-47.
10. E. Sachau, 2003, *Alberuni's India* (Translation of Al-Biruni's *Tahkikat-ul Hind*), New Delhi: Indialog Publications, p. 243.
11. Sachau, op. cit., p. 244.
12. Ahmad, op. cit., p. 147.
13. Ibid., p. 146.
14. Ibid., p. 146.
15. Ibid., pp. 146-47.

the matter dissolves into fundamental particles, every object losing all differentiation.[16]

In light of these facts, the above *Vaiśeṣika-Sūtra*'s meaning is clear.]

karma karma-sādhyam na vidyate |

One *karma* (energy) is not produced by other *karma* (energy).
— *Vaiśeṣika-Sūtras*, 1.1.11)

[An energy needs to do work, and that work in turn gets stored as energy again. Thus there is a cycle of energy-work-energy-work . . . and so on. An energy cannot change into another form of energy without intervening work.]

kārya-virodhi-karma |

Potential energy is against (or opposite) the work.
— *Vaiśeṣika-Sūtras*, 1.1.14

[The accumulated *karma* (potential energy) is exhausted by same amount of work in the opposite direction. Moreover, as potential energy increases, it opposes any further work. Thus in an elastic spring, as potential energy of spring increases, greater is the force opposing any further work. Thus grossly it can be said that *karma* is opposite of or against work.]

Now we will see some *sūtra*s from Chapter 5 of the *Vaiśeṣika-Sūtras*:

ātma-saṁyoga-prayatnābhyam haste karma |

Link with individual (*ātmā*) and (his) effort produces *karma* (energy) in hand. — 5.1.1

tatha-hasta-saṁyogāt-ca-mūsale-karma |

Association with hand (transfers) *karma* in the pulveriser.
— 5.1.2

[When hand lifts a pulveriser (*mūsal*), although "work" is done by the hand, yet the pulveriser develops potential

16. Ahmad, op. cit., p. 147.

energy, when it goes up. This example illustrates that "work" done by one on another object can produce energy in the other object.]

abhighātaje mūṣaladau karmāṇi vyatirekāt-akāraṇam hasta-saṁyogaḥ |

(The pulveriser hits the base, and rebounds. This is a "collision" or *abhighāta*.) The energy generated in the pulveriser on account of the "collision" is not caused due to contact with hand. — 5.1.3

tathā-ātmā-saṁyogo-hasta-karmāṇi |

And because of same reasons, the link with self is not the cause of (generation of) energy in the hand (following the collision). — 5.1.4

[Here the author describes a perfectly elastic collision or recoil of pulveriser with the base. The pulveriser rebounds while the hand is holding it passively. The author is aware that when the pulveriser hits the base, this action has an equal and opposite reaction. This becomes clear from the following *sūtra*s which imply that the pulveriser rises to its original height without any effort by holding hand.]

abhighātāt-mūsala-saṁyogāt-haste karma.

(Thus there are two causes of generation of energy in the hand this time.) The collision (of pulveriser with the base) and (hand's) contact with pulveriser (produce) energy in the hand. — 5.1.5

ātma-karma-hasta-saṁyogāt-ca.

(In this way,) energy (generated in hand may be either due to) the hand's link with the self (or due to its link with an object). — 5.1.6

[Something cannot have movement by its own. The force needs to be exerted on it from outside. The hand in this example, gets motion and thereby energy by different agents external to "hand": either by the "self", i.e. force applied by the individual or by force applied on it by rebounding pulveriser.]

saṁyoga-abhāve gurutvāt patanam |

If the hold is lost, object falls because of gravity. — 5.1.7

[Gravity is a force constantly operating on everything. Anything which is lost from the hold, falls on the earth because of the force of gravity.]

nodana-viśeṣa-abhāvat-na-ūrdhvam na tiryag-gamanam |

(The projectile) cannot move upwards or diagonally up, unless there is (applied) a specific (critical or particular) "impulse". — 5.1.8

[Here we come across the concept of *nodana*. Monier-William's *Sanskrit-English Dictionary* defines *nodana* as "impulse" or driving away. Impulse in today's physics means sudden change of momentum (in a short time). Kaṇāda proposes that a critical "impulse" is required to lift or throw anything upwards.]

prayatna-viśeṣāt nodana viśeṣaḥ |

A particular *prayatna* (force × time) produces a particular *nodana* (change of momentum). — 5.1.9

[*Prayatna* means effort of the individual. Total effort is force applied by a person multiplied by time. The author postulates that there is a fixed relationship (proportion) between this effort by the individual and the amount of "impulse" generated in the object. This *sūtra* anticipates Newton's Second Law of Motion and gives us relationship between force and momentum. Rate of change of momentum (i.e. rate of impulse formation) is proportional to rate of *prayatna*, if we divide both *prayatna* and *nodana* by time which is the same for the two.]

nodana-viśeṣāt-udasana-viśeṣaḥ |

A particular "impulse" (meaning momentum) produces a particular rise (of projectile). — 5.1.10

[Thus if the body is at rest in the beginning, the momentum attained by it is equal to the *nodana* (or impulse) applied to it. For a body of given mass, a particular momentum produces a particular velocity. A particular initial velocity produces a particular maximum height attained by the projectile. It can be said thus that a particular impulse produces a particular height in a given projectile.]

hasta-karmaṇā-dāra-karma-vyākhyātam |

The work done by hand exemplifies "ladies' act" (passive work). — 5.1.11

[When the hand (holding the pulveriser) falls freely with (the pulveriser), it is doing a *dāra-karma* or ladies' act (meaning passive work). Because it is moving in the direction of (gravitational) force. We here get the concept of passive work. *Vaiśeṣika-Sūtras* 5.1.12 to 5.1.15 give further examples of passive work.

If someone explodes a bomb, an innocent person gets burnt. This is an example how someone get an effect passively for others' acts (*VS*, 5.1.12). A sleeping person can walk without doing any effort (*VS*, 5.1.13). The leaves of grass are lifted by wind, when the leaves are lying passively (*VS*, 5.1.14). A magnetic compass moves when a magnet crosses over it. This work is without any visible cause and is an example of passive work (*VS*, 5.1.15).]

There is mention of consecutive decrease in the *saṁskāra* of a vertically thrown projectile. The logic used anticipates the calculus method for calculating the potential and kinetic energy of the projectile.

īśava-yugapat saṁyoga-viśeṣaḥ karma-anyatve hetuḥ
— *VS*, 5.1.16

> When an arrow is shot, it has different (kinetic) energy at different points of the route.
>
> *nodnāt-ādyam-īśoḥ karma tat-karmakāritāt-ca saṁskārat-uttaram tathottaramuttaram ca.*
> *saṁsakāra-abhāve gurutvāt patanam.* — *VS*, 5.1.17-18
>
> Initially the force (*nodana*) applied to the arrow produces initial *karma* (kinetic energy) in the arrow. That energy produces a momentum in the arrow. It is because of this momentum that the arrow goes on going up. As the projectile rises, higher and higher *saṁskāra* becomes less and less. When the *saṁskāra* of the projectile is exhausted, it starts falling down under the influence of gravity.

This is the modern analytical way of studying physics. This type of physics did not exist in Europe before the seventeenth century. Here we also find a good understanding of "reaction to action" when the pulveriser falls at the *ūkhala* and rebounds up. Concept of passive work is also there in the text. The amount of potential energy is equal to the work was also understood which is clear by the *sūtras*, "*kārya virodhi karma*" (Potential energy is against the work; *VS*, 1.1.14). Understanding that a critical minimum force is a must for upward projection of projectile is found in *sūtra* 5.1.8. The *sūtra* tells it in context of gravity, which implies that the force of projection must exceed the force of gravity.

Praśastapāda

About CE 600, Praśastapāda, another Vaiśeṣika scholar of repute, wrote the *Padārthadharma-Saṁgraha* (literally "collection of properties of matter") and commentary on the *Vaiśeṣika Darśana*, the *Praśastapāda Bhāṣya*. These texts are comprehensive books in physics. A wide field ranging from general physics up to quantum physics has been covered. Praśastapāda discusses properties of motion in these. The peculiarity of a single motion affecting a single body at a time, instantaneous velocity, velocity due to gravity, addition of two velocities in opposing direction, vectorial (*digviśiṣṭa*)[17] representation of velocity, etc. have been analysed at least 1,000 years before these concepts were thought of in the West.[18] *Saṁskāras* (previous actions or work done on an object) result in a persistent tendency to move, which is called *vega* (momentum, Newton's First Law).[19] Udyotakara of about the same time, in his *Nyāyavārtika* states that a heavier body falls to ground with greater *vega* than a lighter body (falling from the same height). That means *vega* is a combined function of velocity and mass.

The ancient Indian physicists held that motion is the instantaneous change of place of a particle. They held that a velocity cannot give birth to a new motion, meaning the velocity tends to remain the same and it cannot increase by its own. Impact or push can cause a motion, and the effects of impact or push on the body are called *saṁskāra*. *Vega* is that tendency of a moving body by which it continues its motion. *Vega* is *saṁskāra* has been clearly stated by Praśastapāda. He says that *saṁskāra* is of three types, viz. *vega*, *bhāvanā* (impression) and form-restoring (*sthitisthāpaka*).[20] I have revised my view, and now I feel that *saṁskāra* is momentum and *vega* is its variety. Sen also holds this view.[21] "The power of *saṁskāra* diminishes by doing work (*kāryakāraṇāt*) against countering forces and when the *saṁskāra* is in this way exhausted the moving body comes to rest." Although this description fits both kinetic energy as well as momentum, Chatterjee considers it to be a description of gradual loss of momentum

17. *Utprekṣānādikarma Prakaraṇam* in *Karma-Padārtha-Nirūpaṇam* in *Praśastapāda Bhāṣyam*, Hindi tr. Dhundhiraj Shashtri, Varanasi: Chaukhamba Sanskrit Sansthan, reprint, 2002.

18. S.N. Sen, 1966, "The Impetus Theory of the *Vaiśeṣikas*, *Indian Journal of History of Science*, vol. 1, p. 37.

19. B.N. Seal, op. cit., p. 129.

20. *Saṁskāra Prakaraṇam* in the *Praśastapāda Bhāṣyam*, Hindi tr., p. 221-23.

21. S.N. Sen, op. cit., p. 39-41.

by negative acceleration due to working against force, and hence an anticipation of the equation of force in the Newton's Second Law of Force.[22] But *bhavana* and *sthitisthāpaka* are other two varieties of *saṁskāra*. These I am unable to explain. *Bhāvanā* could be "inertia". *Sthitisthāpaka* literally means form-restoring. It is likely that it meant momentum-equivalent in case of simple harmonic motions or elasticity.

Praśastapāda analyses the velocity due to gravity in a freely falling body in quite modern way. He says that the motion of bodies falling from rest starts by gravity alone. This leads to a *saṁskāra*, in the same direction. As the force of gravity continues to operate, the further motion of the falling bodies is due to gravity as well as *saṁskāra*. Yet further, to this new velocity is added further velocity due to further action of gravity. Thus the velocity goes on increasing. The resultant motion is one but both the causes must be conceived as contributing to the resultant motion.[23] According to Seal, it was a good foundation for explanation of accelerated motion in falling bodies much before anywhere else in the world.[24] The causes of motion have been enumerated as gravity (*gurutva*), pressure or flow of fluids (*dravatva*), effort (*prayatna*) and special contacts (*saṁyogajatva*). Examples of special contact included impact or collision (*abhighāta*) and forced push (*nodana*).[25] In another section, he discusses how energy from hand is transferred to the bow, which transfers it to the arrow, which then comes into motion.[26]

---

22. S.D. Chatterjee, 2004, "Physics and Mechanics in Ancient and Medieval India", in *The Cultural Heritage of India*, vol. IV, ed. P. Ray and S.N. Sen, Kolkata: The Ramakrishna Institute of Culture, p. 104.
23. *Apratyayakarma Prakaraṇam* in the *Praśastapāda Bhāṣyam*, Hindi tr.
24. B.N. Seal, 1958, op. cit., p. 141.
25. *Apratyayakarma Prakaraṇam* in the *Praśastapāda Bhāṣyam*. Also see S.D. Chatterjee, op. cit., p. 103.
26. *Satya-pratyaya-karma Prakaraṇa* in *Praśastapāda Bhāṣyam*, Hindi tr., pp. 261-62.

Five types of simple movements have been noted, vertically upward throw (*utkṣepaṇa)*, vertically downward fall or throw (*apakṣepaṇa*), attraction (*ākuñcana*), repulsion (*prasāraṇa*) and motion (*gamana*).[27] Cause of translatory motions is new contact with a point in space then separation with that point in the next short interval of time and a new contact with the next point in space-field (*dik-pradeśa-saṁyoga-vibhāga-kāraṇam*).[28] This precisely gives us understanding of the modern concept of *dx/dt* for velocity. Praśastapāda mentions that *karma* (energy) is instantaneous (*kṣaṇikatvam*), belongs to material substances (*mūrtta-dravya-vr̥ttitvam*), is finished on doing work by itself (*sva-kārya-saṁyoga-virodhitvam*), can initiate work on association with some other object (*sva-paraśraya-samaveta-kāryarambhakatvam*), is basis of various types of actions (*pratiniyata-jāti-yogitvam*), cannot form material substances (*dravya-anārambhakatvam*), causes motion in a specific direction (vector) (*digviśiṣṭa-kāryārambhakatvam*).[29] He gives the various varieties of motion such as, curvilinear motion or speed (*gamana*), rotatory motion (*bhramaṇa*) and simple harmonic motion or vibratory motion (*spandana).*[30] In the discussion of circular motion, he also refers to angular momentum (*avayavinaḥ saṁskārāt*).[31]

Praśastapāda presents the ancient Indian concept of the theory of relativity, where relativity has been called *apekṣā-buddhi.* It is a cause of relative appearance of time and space and its destruction leads to dissociation between space, time and matter/energy.[32] Time is something eternal, it is an eternal

27. *Gamana Prakaraṇa* in *Praśastapāda Bhāṣyam,* Hindi tr. pp. 245-46.
28. *Utkṣepaṇādikarma Prakaraṇam* in *Praśastapāda Bhāṣyam,* Hindi tr., p. 244.
29. Ibid., p. 240.
30. S.D. Chatterjee, op. cit., p. 103.
31. *Apratyayakarma Prakaraṇam* in the *Praśastapāda Bhāṣyam,* Hindi tr., p. 269.
32. *Paratvaparatva Prakaraṇam,* in the *Praśastapāda Bhāṣyam,* Hindi tr., pp. 124-36 (part 2).

dimension of existence. But it is relative also and its perception is dependent on the conditions of the subject perceiving it. Many of the features of time are like that of space. It is *ādhāra* of events in the same way as space is *ādhāra* of objects. Time is *anāśrita* (not dependent, not requiring others). On that account also it is very much like space. It is also suggested that in the very beginning of universe, it was a condensed form of time from which everything emerged.[33]

It is remarkable to note that the word *gurutva* was used to denote weight (as an indicator of mass) as well as the force of gravity. Today in contemporary physics also, the same practice loosely exists.[34] Praśastapāda writes that gravity (*gurutva*) is the cause of fall of solids and liquids and it can be prevented by contact and force as well as (upward) momentum.[35] He was intelligent enough to define matter by its capacity to gain momentum by application of force (*yat-adravyam tat kriyavanna bhavati*).[36]

---

33. *Kāla Prakaraṇam*, in the *Praśastapāda Bhāṣyam*, Hindi tr., pp. 41-45.
34. S.D. Chatterjee, op. cit., p. 104.
35. *Gurutva Prakaraṇam*, in the *Praśastapāda Bhāṣyam*, Hindi tr., pp. 217-18 (part 2).
36. *Praśastapāda Bhāṣya*, (*Nidarśana-prakaraṇam*), p. 197.

PART - III
MATHEMATICAL SCIENCES

8

Numbers and Zero in Vedic Sūktas
Some Applications and China Contacts

*Bhu Dev Sharma**

Languages and mathematics make human expression possible and exact. While languages vary, mathematics in its development is universally the same. Where were the basics of mathematics realised and developed? In the paper we trace the roots of numbers and geometry in Vedas. Internal evidence from Vedas is produced about numbers 1, 2, . . . , 9 and zero being there. The Vedas also refer to what in the modern terminology are called "sequences of numbers" and fractions, both unit and others. Some applications in the form of special branches covered by Upavedas and in war and communication are pointed out. While travel of Hindu mathematics to West is widely known, here we bring mention of the presence of Indian mathematicians in China. Lastly, brief mentions of great Indian mathematicians, starting from Baudhāyana (800 BCE); to Jagannatha (CE 1750) are made to indicate the continuity of the tradition.

Some Quotes

> Two hundred years before Pythagoras, philosophers in northern India had understood that gravitation held the solar system together, and that therefore the sun, the most massive object, had to be at its center.
>
> — Dick Teresi, *Lost Discoveries*, 2002, p. 1

---

* Professor of Mathematics, JIIT University, Noida - 201 307 (UP) <bhudev_Sharma@yahoo.com>

Twenty-four centuries before Isaac Newton, the Hindu *Ṛgveda* asserted that gravitation held the universe together, . . . The Sanskrit-speaking Āryans subscribed to the idea of a spherical earth in an era when the Greeks believed in a flat one. The Indians of the fifth century CE somehow calculated the age of the earth as 4.3 billion years; scientists in nineteenth-century England were convinced it was 100 million years. The modern estimate is 4.6 billion years.

— Ibid., pp. 7-8

For some, the failure to acknowledge the success of non-Western cultures derives not just from ignorance, but from a conspiracy.

— Ibid., p. 11

The zero, to be sure, first liberated the digits from the counting board and enabled them to stand alone. . . . With the adoption of the zero, India finally attained the abstract place-value notation which was about to begin its journey of conquest through the world as the most mature and highly developed form of numerals.

— Karl Menninger, *Number Words and Number Symbols: A Cultural History of Numbers*, 1969

The importance of the creation of zero mark can never be exaggerated. This giving to airy nothing, not merely a local habitation and a name, a picture, a symbol, but helpful power, is the characteristic of the Hindu race, whence it sprang. It is like coining the nirvana into dynamos. No single mathematical creation has been more potent for the general on-go of intelligence and power."

— G.B. Halsted, *On the Foundation and Technique of Mathematics*

How important was mathematics considered, in India, for understanding anything in the world?

बहुभिर्प्रलापैः किम्, त्रयलोके सचराचरे।
यद् किंचिद् वस्तु तत्सर्वम्, गणितेन बिना न हि।।

bahubhirpralāpaiḥ kim, trayaloke sacarācare।
yad kiñcid vastu tatsarvam, gaṇitena binā na hi।।

— Mahāvīrāchārya, *Gaṇita Sāra-Saṁgraha*, 9th century

Why cry profusely, in this threefold world, if there exists something, live or inert, has to have its mathematics (cannot be understood without mathematics for it).

Introduction

MATHEMATICS, it can be readily accepted by everyone, is an intellectual discipline. It is also, along with a language, the most basic form of expression for human dealings and interaction. This is the reason that some kind of number system is found in every ancient civilisation. It should astonish a thinking mind, that while many languages exist and flourish in the world, only the decimal system is universally used.

Just to make the last point clear, we do find that two systems currently in vogue. These are the so-called Roman System (using symbols I, II, III, IV, V, . . . , X, . . ., C, . . ., M, . . . for 1, 2, 3, . . . etc.) and commonly called the "decimal system" (using ten symbols, viz. 1, 2, 3, 4, 5, 6, 7, 8, 9, 0). It is common knowledge (sometimes not realised though), that no mathematical operation, not even the simplest like adding two numbers to give a third one, like IV + VI = X, can be defined in the Roman System. This knocks out Roman System for any meaningful purpose in the scientific world.

On the other hand, scientists of any generation wonder on the perfect nature of *the place value system* — the decimal system. This is perhaps the only perfect system in the scientific world ever found. All historians of science and mathematics give the credit of developing this system to Hindus of the ancient. However, due to limitations of the scientists having no exposure to Vedas, the most ancient of the world heritage, the roots of these are not clearly traced from there.

The paper provides internal evidence from Vedas that the numbers 1 to 9 and the concept of "zero" and the "place-value notion" were known to Vedic people. Going to the very roots, presenting direct internal evidence from Vedas, the paper presents the following points:

1. The numbers 1, 2, 3, . . . 9, with 9 as the largest single digit find mention in the *ṛcā*s and *mantra*s of the Vedas;
2. There is enough evidence that a method to denote "zero" was known to Vedic seers;
3. The Vedas refer to what in the modern terminology are called "sequences of numbers";
4. That the idea of fractions, both unit and others, has also found clear mention in Vedas. It further points out how initially proposed theories by European scholars of mathematics, considering algebra and geometry of Greek and/or Old-Babylonian origin have come to be discarded, based on evidence produced from Vedic works like *Śutapatha Brāhmaṇa* and *Taittirīya Saṁhitā*.
5. The paper also brings out some references of *Indian mathematicians spending time in China in eighth century and using Indian mathematics there for calendrical purposes.*

Evolving Theories of Origin of Mathematics in India rather than in Greece and Babylonia

Mathematics originated in numbers and geometrical figures. The European renaissance of 1300–1600 traced roots of everything to Ionian Greece, in particular to sixth century BCE. Following these lines, W.W.R. Hall in 1901 wrote, "The history of Mathematics cannot with certainty be traced back to any school or period before that of Ionian Greeks." In these early years of European intellectual growth, contributions of India were obviously not considered. After *Sanskrit studies attracted some European scholars, this situation changed.* In 1875 G. Thibaut, a Sanskrit scholar with a view to informing the learned world about Indian mathematics, translated a large part of *Śulbasūtras.* In 1877 Cantor, realising the importance of Thibaut's work, began a comparative study of Greek and Indian mathematics.

Initially he concluded that Indian geometry is derived from Alexandrian knowledge. However, some 25 years later, with greater study, he concluded that the Indian geometry and Greek geometry are related. The process of assigning dates was also picked up. And later, Cantor eventually conceded a much earlier date to Indian geometry.

There then came another turn. In 1928 Neugebauer published a paper in which he traced that the so-called Pythagoras theorem was known well over a thousand years before Pythagoras, but in 1937 made a hazard guess of geometry being of Babylonian origin. Seidenberg (1978) expanded the study of Vedic sources, including *Śatapatha Brāhmaṇa* and *Taittirīya Saṁhitā* closely comparing Greek and Vedic Mathematics as well as "Old-Babylonian and Vedic Mathematics." He concluded as follows:

> . . . geometric algebra existed in India before the classical period of Greece. A comparison of Pythagorean and Vedic mathematics together with some chronological considerations showed . . . [that] a common source for Pythagorean and Vedic mathematics is to be sought either in Vedic mathematics or in an older mathematics much like it. The view that Vedic mathematics is a derivative of Old-Babylonian [is] rejected.

Number Evidence Directly from Vedas

In tracing the origin of the present-day universal ideas about numbers, our search begins from Vedas. The famous dictum of *Ṛgveda, ekam sad viprā bahudhā vadanti* uses *ekam* meaning "one" as the cardinal number. Further to search numbers in Vedas, let us refer to the famous and fundamental stanza — *Puruṣa-Sūkta* — of *Ṛgveda*:

सहस्रशीर्षा पुरुषः सहस्राक्षः सहस्रपात्।
स भूमिं विश्वतो वृत्वा अत्यतिष्ठ**द्दशा**ङ्गुलम्।।1।।

sahasra*śīrṣā puruṣaḥ* **sahas***rākṣaḥ* ***saha****srapāt* ।
sa bhūmiṁ viśvato vṛtvā 'tyatiṣṭhaddaśāṅgulam ॥1॥

पुरुष एवेदं सर्वम् यद्भूतं यच्च भव्यम्।
उतामृतत्वस्येशानो यदन्नेनातिरोहति॥२॥

puruṣa evedaṁ sarvam yadbhūtaṁ yacca bhavyam ।
utāmṛtatvasyeśānoḥ yadannenātirohati ॥2॥

एतावानस्य महिमातो ज्यायाँश्च पूरुषः।
पादोऽस्य विश्वा भूतानि, **त्रि**पादस्यामृतं दिवि॥३॥

etāvānasya mahimāto jyāyāṁśca pūruṣaḥ ।
pādo 'sya viśvā bhūtāni ***tri****pādasyāmṛtaṁ divi* ॥3॥

त्रिपादूर्ध्वं उदैत्पुरुषः पादोऽस्येहाभवत् पुनः।
ततो विष्वङ्व्यक्रामत्, साशनानशने अभि॥४॥

tri*pādūrdhvaṁ udaitpuruṣaḥ pādo 'syehābhavat punaḥ*।
tato viṣvaṅvyakrāmat sāśanānaśane abhi ॥4॥

तस्माद् विराडजायत, विराजो अधि पूरुषः।
स जातो अत्यरिच्यत, पश्चाद् भूमिमथो पुरः॥५॥

tasmād virāḍajāyata virājo adhi pūruṣaḥ ।
sa jāto atyaricyata paścād bhūmimatho puraḥ ॥5॥

यत्पुरुषेण हविषा, देवा यज्ञमतन्वत।
वसन्तो अस्यासीदाज्यम्, ग्रीष्म इध्मश्शरद्धविः॥६॥

yat puruṣeṇa haviṣā, devā yajñamatanvata ।
vasanto asyāsīdājyam, grīṣma idhmaś śaraddhaviḥ ॥6॥

सप्तास्यासन्परिधयः **त्रि सप्त** समिधः कृताः।
देवा यद् यज्ञं तन्वानाः, अब्ध्नन्पुरुषं पशुम्॥१५॥

saptāsyāsanparidhayaḥ **tri sapta** *samidaḥ kṛtāḥ*।
devā yad yajñaṁ tanvānāḥ, abdhnanpuruṣaṁ paśum ॥15॥

तं यज्ञं बर्हिषि प्रौक्षन्, पुरुषं जातमग्रतः।
तेन देवा अयजन्त, साध्या ऋषयश्च ये॥७॥

taṁ yajñaṁ barhiṣi praukṣan puruṣaṁ jātamagrataḥ ।
tena devā ayajanta, sādhyā ṛṣayaśca ye ।।7।।
— *Ṛgveda*, X.90.1-6, 15, 7

Here, look at bold words, one clearly finds use of the words *sahasra* (thousand), *daśa* (ten), *tri* (three), *sapta* (seven), etc. for numbers.

ATHARVAVEDA SŪKTA ON NUMBERS

Besides sporadic use of numbers in Vedic texts, there is a complete *sūktam* in *Atharvaveda* that is devoted to numbers from 1 to 11, with all the names — *eka, dvī, tri, catura, pañca, ṣaḍa, sapta, aṣṭa, nava, daśa, ekādaśa* — in proper order. These names of numbers are practically intact to the present day. The use of base ten is also very clear in this *sūktam* when 11 is referred to *ekādaśa*. The *sūktam* is as follows:

यद्येक वृषोऽसि सृजारसोऽसि। यदि द्विवृषोऽसि सृजारसोऽसि।।१।।
यदि त्रिवृषोऽसि सृजारसोऽसि। यदि चतुर्वृषोऽसि सृजारसोऽसि।।२।।
यदि पञ्चवृषोऽसि सृजारसोऽसि। यदि षड्वृषोऽसि सृजारसोऽसि।।३।।
यदि स्प्तवृषोऽसि सृजारसोऽसि। यद् अष्टवृषोऽसि सृजारसोऽसि।।४।।
यदि नववृषोऽसि सृजारसोऽसि। यदि दशवृषोऽसि सृजारसोऽसि।।५।।
यद् एकादशोऽसि सोऽपोदकोऽसि।
– अथर्ववेद पञ्चम काण्डम्, चतुर्थोऽनुवाक, सूक्तम् १६

*yad**yeka**vṛṣo 'si sṛjāraso 'si* । *yadi* ***dvi****vṛṣo 'si sṛjāraso 'si* ।।1।।
yadi ***tri****vṛso 'si sṛjāraso 'si* । *yadi* ***catur****vṛṣo 'si sṛjāraso 'si* ।।2।।
yadi ***pañca****vṛṣo 'si sṛjāraso 'si* । *yadi* ***ṣaḍ****vṛṣo 'si sṛjāraso 'si* ।।3।।
yadi ***sapta****vṛṣo 'si sṛjāraso 'si* । *yad* ***yaṣṭa****vṛṣo 'si sṛjāraso 'si* ।।4।।
yadi ***nava****vṛṣo 'si sṛjāraso 'si* । *yadi* ***daśa****vṛṣo 'si sṛjāraso 'si* ।।5।।
yad ***ekādaśo*** *'si so 'podako 'si* । — *Atharvaveda* 5.4.16

It is of some significance to see that the *sūkta* ends after introducing *ekādaśa* for 11, as it is easy to now name 12 as *dvādaśa*, 13 as *trayodaśa* and so on. Thus the *sūkta* introduces the numbers 1 to 9 to the base 10 fully and clearly.

Use of Zero Suggested in Atharvaveda

Next, we come to the very important concept of zero. In a rather implicit manner, we find, elsewhere in *Atharvaveda* again, special place for numbers and their multiples of ten. A complete *sūkta* outlining special use of numbers and their ten-multiples, viz. of 1 and 10, 2 and 20, 3 and 30, 4 and 40, . . ., 10 and 100 is as follows:

एका च मे **दश** च मेऽपवक्तार ओषधे।
ऋतजात ऋतावरि मधु मे मधुला करः।।१।।

द्वे च मे **विंशतिश्च** मेऽपवक्तार ओषधे।
ऋतजात ऋतावरि मधु मे मधुला करः।।२।।

तिस्रश्च मे **त्रिंशच्च** मेऽपवक्तार ओषधे।
ऋतजात ऋतावरि मधु मे मधुला करः।।३।।

चतस्रश्च मे **चत्वारिं**शच्च मेऽपवक्तार ओषधे।
ऋतजात ऋतावरि मधु मे मधुला करः।।४।।

पञ्च च मे **पञ्चाश**च्च मेऽपवक्तार ओषधे।
ऋतजात ऋतावरि मधु मे मधुला करः।।५।।

षट् च मे **षष्टि**श्च्च मेऽपवक्तार ओषधे।
ऋतजात ऋतावरि मधु मे मधुला करः।।६।।

सप्त च मे **सप्तति**श्च्च मेऽपवक्तार ओषधे।
ऋतजात ऋतावरि मधु मे मधुला करः।।७।।

अष्ट च मे **अशीति**श्च्च मेऽपवक्तार ओषधे।
ऋतजात ऋतावरि मधु मे मधुला करः।।८।।

नव च मे **नवति**श्च्च मेऽपवक्तार ओषधे।
ऋतजात ऋतावरि मधु मे मधुला करः।।९।।

दश च मे **शतं** च मेऽपवक्तार ओषधे।
ऋतजात ऋतावरि मधु मे मधुला करः।।१०।।

शतं च मे **सहस्रं** चापवक्तार मेऽपवक्तार ओषधे।
ऋतजात ऋतावरि मधु मे मधुला करः।।११।।
– अथर्ववेद पञ्चम काण्डम्, चतुर्थोऽनुवाक, सूक्तम् १५

ekā *ca me* ***daśa*** *ca me 'pavaktāra oṣadhe* ।
ṛtajāta ṛtāvari madhu me madhulā karaḥ ।।1।।

dve *ca me* ***viṁśatiś****ca me 'pavaktāra oṣadhe* ।
ṛtajāta ṛtāvari madhu me madhulā karaḥ ।।2।।

tri*sraśca me* ***triṁś****ac ca me 'pavaktāra oṣadhe* ।
ṛtajāta ṛtāvari madhu me madhulā karaḥ ।।3।।

catasr*aśca me* ***catvāriṁśa****c ca me 'pavaktāra oṣadhe* ।
ṛtajāta ṛtāvari madhu me madhulā karaḥ ।।4।।

pañca *ca me* ***pañcāś****ac ca me 'pavaktāra oṣadhe* ।
ṛtajāta ṛtāvari madhu me madhulā karaḥ ।।।5।।

ṣaṭ *ca me* ***ṣaṣṭi****śca me 'pavaktāra oṣadhe* ।
ṛtajāta ṛtāvari madhu me madhulā karaḥ ।।6।।

sapta *ca me* ***saptatiś****ca me 'pavaktāra oṣadhe* ।
ṛtajāta ṛtāvari madhu me madhulā karaḥ ।।7।।

aṣṭa *ca me* ***aśītiś****ca me 'pavaktāra oṣadhe* ।
ṛtajāta ṛtāvari madhu me madhulā karaḥ ।।8।।

nava *ca me* ***navatiś****ca me 'pavaktāra oṣadhe* ।
ṛtajāta ṛtāvari madhu me madhulā karaḥ ।।9।।

daśa *ca me* ***śataṁ*** *ca me 'pavaktāra oṣadhe* ।
ṛtajāta ṛtāvari madhu me madhulā karaḥ ।।10।।

śataṁ *ca me* ***sahasraṁ*** *cāpavaktāra oṣadhe* ।
ṛtajāta ṛtāvari madhu me madhulā karaḥ ।।11।।

— *Atharvaveda* 5.4.15

Nine the Largest Single Digit Number: Ṛgveda Evidence

From above references, it can be inferred without doubt that the Vedic seers had a fully developed number system with names and clear evidence of using base 10 for higher numbers. That 9 was taken to be the single largest number is clearly indicated in another *mantra* of *Ṛgveda*. In the following *mantra* addressed to Indra, the use of repeated 9 is to construct a large number. What it says is that like a terrified hawk, you (Indra) have crossed rivers whose number is formed from repeated 9s (meaning many) rivers:

अहेर्यातारं कमश्य इन्द्र हृदि यत्ते जघ्नुषो भीरगच्छत्।
नव च यन्नवतिं च स्रवन्तीः श्येनो न भीतो अतरो रजांसि।।
– ऋग्वेद मंडल १, अध्याय ७, सूक्त ३२, ऋचा १४

aheryātāraṁ kamaśya indra hṛdi yatte jaghnuṣo bhīragacchat ।
nava ca yannavatiṁ *ca sravantīḥ śyeno na bhīto ataro rajāṁsi*।। — *Ṛgveda*, 1.32.14

Symbols for Numbers

As is well known, giving symbols to any concept is a well-recognised Vedic practice. However, Vedas are Śrutis, meaning thereby that these were preserved and passed on by oral tradition. The numbers, naturally, in Vedas, had to be recoded by their names even though symbol representations must have been in use for them. Therefore it is incorrect to infer that Vedic seers did not have symbols for numbers, and that the symbols were not in use during the Vedic times. In fact, the names like *ekādaśa* indicate that numbers greater than 10 were given names that had representation in some symbols, with turning points coming at TEN, TWENTY, . . . NINTY, . . . HUNDRED, etc. This, on the contrary, indicates knowledge and use of all important characters, the place-value system. The *sūkta* devoted to numbers and their 10-multiples brings us, in fact, close to inferring that if a number from 1 to 9 was written in some way, their 10-multiples were written in a certain related way to this base number. This suggests use of some symbol for zero. We cannot expect to get in Vedas those symbols. It brings us to next point that we present.

EVIDENCE THAT NUMBERS WERE WRITTEN

When did the numbers start to be written in some symbol forms? Vedic system being rich in symbolism, the correct guess is — right from the time their concepts developed and matured. The earliest thoughts of India's learned and the elite are represented in *Ṛgveda*, wherein we find references of written numbers.

We like to draw attention to use of dice with six faces and numbers written on them in 10.34.2, where we find:

अक्षस्याहमेकपरस्य हेतो

akṣasyāhameka parasya heto

Here use of word "dice" (*akṣa*), in the game indicates that the faces had some marks — perhaps of numbers — on them. Also another *mantra* from *Ṛgveda* that clearly talks of written numbers is

इन्द्रेण युजा निःसृजन्त वाघतो व्रजं गोमन्तमश्विनम्।
सहस्रं मे ददतो **अष्टकर्ण्यः** श्रवो देवेष्वक्रत।। – ऋग्वेद १०.६२.७

indreṇa yujā niḥsṛjanta vāghato vrajaṁ gomantamaśvinam ।
sahasraṁ *me dadato* ***aṣṭakarṇyaḥ*** *śravo deveṣvakrata* ।।

— *Ṛgveda* 10.62.7

Here it mentioned that "1000 cows, which had figure of 8 written on their ears, were given to me".

In fact, writing numbers, in some form, must have started with any other form of written word if not before.

As we know so well, the tradition of a written language also goes to antiquity in India. Mother of most of the language of the world, Vedic Sanskrit, was once written in Brāhmī script, which is considered to be the precursor of Devanāgarī, and so many other alphabets of India. Another script used in ancient India was Kharoṣṭhi, which unlike Brāhmī was written right to left. It is believed to have originated in the north-west and was in use from the fifth century BCE to third century CE. In terms of physical evidence available today, world's oldest written document that uses present-day numerals 1, 2, . . ., 9, is of 595 CE. It is a piece of "Gurjardeśa". Also the earliest known written zero is found on "Gwalior inscription" of CE 870.

MOHEṄJO-DĀṚO AND AŚOKA WRITINGS

From Moheṅjo-Dāṛo writings, there is evidence of written numbers from 1 to 13. Evidence of written language, in a kind of Brāhmī script, is found on the vessels, estimated to be of 3000-6000 BCE, in Madras Museum.

Most of the stone writings of Emperor Aśoka are in Brāhmī and some in Kharoṣṭhi. Kharoṣṭhi numbers are certainly not the ancestors of the Sanskrit or present numerals. Brāhmī script had symbols for 1, 2, 3, 4, 5, 6, 7, 8, 9, 10, 20, 30, 40, 50, 60, 70, 80, 90, 100, 200, . . . , 1000, 2000, etc. We will not go into linguistic details in the present article. But it is well recognised that there is substantial scholarly work showing that from Sanskrit resulted Greek and Iranian languages in 500 BCE, Latin in 300 BCE, Gothic in CE 400, Celtic, N&W Germanic and Old High German in CE 800, Old Savic in CE 900, Lithuanian in CE 1700, etc.

Nature of Vedic Numbers: Abstract Concepts in Vedic Thought

Mathematics, in its true nature, has been termed abstract — man's mental eye — a train of thoughts. It allows him to roam into a world far remote from his nature he sees around. How and when did the number thoughts started in the Vedic mind?

Most people think that numbers must have started with

counting, but then this would not give rise to discovery of zero, as also not to mystic significance of numbers in Hindu-Vedic tradition. The ten digits 1, 2, . . ., 9, 0, in my opinion, also do not arise from the simplistic thought that there are 10 fingers in two hands, which again suggest counting as the basis of numbers, the digits other than 0 being only nine. This is again due to the fact that 0 then again remains unexplained.

Obviously, Hindu concept of numbers is related to abstract philosophical thoughts. Let me quote again from *Atharvaveda*, a *ṛcā* that mystifies the number one.

तमिदं निगतं सह स एष एक एकवृदेक एव।
य एतं देवमेकवृतं वेद।।

सर्वे अस्मिन् देवा एकवृतो भवन्ति।
य एतं देवमेकवृतं वेद।। – अथर्ववेद १३.४.१२-१३

tamidaṁ nigataṁ saha sa eṣa eka ekavṛdeka eva ।
ya etaṁ devamekavṛtaṁ veda ।।

sarve asmin devā ekavṛto bhavanti ।
ya etaṁ devamekavṛtaṁ veda ।।
— *Atharvaveda*, 13.4.12-13

He has that power. He is this ONE. There is only ONE with the strength of ONE. All *deva*s share that strength of ONE. Translated in mathematical terms, this *ṛcā* would mean — ONE multiplied with ONE, remains ONE, and that ONE is always a factor of any quantity.

The place-value system also must have got its stimulus by the Hindu social system in which people of higher rank sit in a higher place than those junior to them in an ordered way.

Series and Sequences in Vedas

Interestingly, there is enough internal evidence in Vedas, that mathematical patterns that, in today's language, may be termed as "series and sequences", were considered an important part of Vedic knowledge. It may be mentioned that sets and sequences are the very basic concepts and structures in pure mathematics.

Chapter 18 of the *Yajurveda*, which forms what is called the *Vājasaneyī Saṁhitā*, is considered important from the point of view of knowing a vast spectrum of religious obligations, agriculture and agricultural produce, domestication of animals, live and inert things, objects of worship, duties and behaviour of men, duties and behaviour of a ruler. This part deals with concepts and ideas, both worldly and spiritual, that prevailed during the Vedic Saṁhitā period. Also, in this chapter names of all domestic animals, all different types of grains are listed and there seems to be instructions to obtain them through *yajña*. Mathematical concepts about sequences appear here in two stanzas, as follows:

एका च मे तिस्रश्च मे
पञ्च च मे सप्त च मे
नव च मऽएकादश च
त्रयोदश च मे पञ्चदश च मे
सप्तदश च मे नवदश च मे
ऽएकविंशतिश्च मे त्रयोविंशतिश्च मे
पञ्चविंशतिश्च मे सप्तविंशतिश्च मे
नवविंशतिश्च मऽएकत्रिंविंशतिश्च

तिस्रश्च मे पञ्च च मे
सप्त च मे नव च
मऽएकादश च मे त्रयोदश च मे
पञ्चदश च मे सप्तदश च मे
नवदश च मऽएकविंशतिश्च मे
त्रयोविंशतिश्च मे पञ्चविंशतिश्च मे
सप्तविंशतिश्च मे नवविंशतिश्च मे
मऽएकत्रिविंशतिश्च मे यज्ञेन कल्पताम्॥

— यजुर्वेद १८.२४

ekā *ca me* ***tisraśca*** *me*
pañca *ca me* ***sapta*** *ca me*
nava *ca me* ***ekādaśa*** *ca me*
trayodaśa *ca me* ***pañcadaśa*** *ca me*
saptadaśa *ca me* ***navadaśa*** *ca me*
ekaviṁśatiśca *me* ***trayoviṁśatiśca*** *me*
pañcaviṁśatiśca *me* ***saptaviṁśatiśca*** *me*
navaviṁśatiśca *ma* ***ekatriṁśaśca***
ma ***ekatriṁśaśca*** *me yajñena kalpatām* ||

tisraśca *me* ***pañca*** *ca me*
sapta *ca me* ***nava*** *ca me*
ekādaśa *ca me* ***trayodaśa*** *ca me*
pañcadaśa *ca me* ***saptadaśa*** *ca me*
navadaśa *ca me* ***ekaviṁśatiśca*** *me*
trayoviṁśatiśca *me* ***pañcaviṁśatiśca*** *me*
saptaviṁśatiśca *me* ***navaviṁśatiśca*** *me*

— *Yajurveda*, 18.24

The sequence of pair of numbers given here is (1, 3), (3, 5), (5, 7), (7, 9), (9, 11), (11, 13), (13, 15), (15, 17), (17, 19), (19, 21), (21, 23), (23, 25), (25, 27), (27, 29), (29, 31). These are obviously pairs of odd numbers. By giving them in pairs, it seems that several patterns arising by their pairwise sum, difference and product, etc. must have been explained by the teacher.

The pattern arising by sum of these pairs is; 4, 8, 12, 16, 20, 24, 28, 32, 36, 40, 44, 48, 52, 56, 60, ...

Here the terms are progressing by the same number 4. While the pattern arising by their product is 3, 15, 35, 63, 99, 143, 195, . . . , where the numbers are progressing with variable difference pattern 12, 20, 28, 36, ...

Yet another sequence of paired numbers found in *Yajurveda* is:

चतस्रश्च मेऽष्टौ च मे ऽष्टौ च मे द्वादश च मे
द्वादश च मे षोडश च मे षोडश च मे विंशतिश्च मे
विंशतिश्च मे चतुर्विंशतिश्च मे चतुर्विंशतिश्च मे ऽष्टाविंशतिश्च मे
ऽष्टाविंशतिश्च मे द्वात्रिंशतिश्च मे द्वात्रिंशतिच्च मे षट्त्रिंशतिच्च मे
षट्त्रिंशतिच्च मे चत्वात्रिंशतिच्च मे
चत्वारिंशतिच्च मे चतुश्चत्वारिंशतिच्च मे
चतुश्चत्वारिंशतिच्च मेऽष्टाचत्वारिंशतिच्च मे यज्ञेन कल्पताम्।।२५।।

catasraśca me ***aṣṭau*** *ca me* ***aṣṭau*** *ca me* ***dvādaśa*** *ca me*
dvādaśa *ca me* ***ṣoḍaśa*** *ca me* ***śoḍaśa*** *ca me* ***viṁśatiśca*** *me*
viṁśatiśca *me* ***caturviṁśatiśca*** *me* ***caturviṁśatiśca*** *me* ***aṣṭāviṁśatiśca*** *me*
aṣṭāviṁśatiśca *me* ***dvātriṁśatiśca*** *me* ***dvātriṁśaticca*** *me* ***ṣaṭtriṁśaticca*** *me*
ṣaṭtriṁśaticca *me* ***catvāriṁśaticca*** *me* ***catvāriṁśaticca*** *me* ***catuścatvāriṁśaticca*** *me*
catuścatvāriṁśatticca *me* ***aṣṭcatvāriṁśaticca*** *me yajñena kalpatām*॥

— *Yajurveda*, 18.25

The sequence of pair of numbers given here is (4, 8); (8, 12), (12, 16), (16, 20), (20, 24), (24, 28), (28, 32), (32, 36), (36, 40), (40, 44), (44, 48), . . . which must have been selected for its many properties not listed. It may be mentioned that unless these sequences had some very interesting properties, these will not find place in the Veda. It was a method of Vedas, perhaps to encourage creativity and spirit of inquiry that details were, in general, left out.

Indian Mathematicians in China

Another interesting study has recently been brought out by Nobel Laureate Amartya Sen (*The Argumentative Indian,* 2005). Sen writes:

> Calendrical studies, in which Indian astronomers located in China in the eighth century, . . . were particularly involved, made good use of the progress of trigonometry that had already occurred in India by then (going much beyond the original Greek roots of Indian trigonometry).
>
> The movement east of Indian trigonometry to China was part of a global exchange of ideas that also went west around that time.
>
> Indeed this was also about the time when Indian trigonometry was having a major impact on the Arab world (with widely used Arabic translations of the works of Āryabhaṭṭa, Varāhamihira, Brahmagupta and others) which would later influence European mathematics as well, through the Arabs.

In a chapter on "China and India", Sen mentions,

> Several Indian mathematicians and astronomers held positions in China's scientific establishment, and an Indian scientist, Gautama Siddhārtha (Qutan Xida, in Chinese) even became the president of the official Board of Astronomy in China in the eighth century.

Sen further points out,

> Gautama (Qutan Xida) produced the great Chinese compendium of astronomy Kaiyvan Zhanjing — an eighth-century scientific classic. He was also engaged in adopting a number of Indian astronomical works into Chinese. For example, Jiuzhi Li, which draws on a particular planetary

> calendar in India (Navagraha calendar) is clearly based on the classical *Pañcasiddhāntika* . . . , produced around CE 550 by Varāhamihira.
>
> It is mainly an algorithmic guide to computation, estimating such things as the duration of eclipses based on the diameter of the moon and other relevant parameters. The techniques involved drew on methods that were established by Āryabhaṭṭa and then further developed by his followers in India such as Varāhamihira and Brahmagupta.

Applications of Mathematics and Sciences

Mathematics, as we know, has applications ranging from day-to-day situations to travelling to stars. Applications of science (knowledge applicable to physical world), in Vedic times have been divided into Upaveda. These corresponding to each four Vedas are: Āyurveda, Dhanurveda, Sthāpatyaveda, Ghandharvaveda.

These are, in my understanding, the Sanskrit works of applied Sciences of the Vedic Tradition.

IN WAR AND COMMUNICATION

There are mentions of fierce wars recorded in Sanskrit texts. These mention different types of weapons, some seem to be atomic and powerful other devices. Another area is about means of travel. While we do not find mention of motorised vehicles, which are very inconvenient, need a lot of road developments, the wisdom of those times, seem to have devised various types of aeroplanes. Whose mentions are rampant.

Śulbasūtras

Śulba means "rope" and *sūtra* means "formula". Thus *Śulbasūtra* literally means "formulas of measurements" which basically means "geometry". Their origin can be traced to Vedas and scholars consider that these must have been known at least eight centuries BCE. The ideas presented therein seem to be motivated by considerations of forming *yajña* altars. *Śulbasūtra* contains what is called Pythagoras theorem, but not the proof.

BAUDHĀYANA

Two main *Śulbasūtras* are that of Baudhāyana and of Āpastamba. Bodhāyana lived around 800 BCE and Āpastamba 600 BCE. *Baudhāyana Śulbasūtra* contains geometric solutions (not algebraic) of a linear equation in a single unknown. Quadratic equations of the form $ax^2 = c$ and $ax^2 + bx = c$ also appear. Several values of *pi* occur in Bodhāyana's *Śulbasūtra*. Bodhāyana uses different approximations for constructing circular shapes. An interesting and correct to four decimal places value of irrational number *square root of* 2 is given in the form = 1 + 1/3 + 1/(3×4) – 1/(3×4×34).

ĀPASTAMBA SŪTRA

Āpastamba Sūtra is a relatively more extensive work having six chapters, while *Baudhāyana Sūtra* had three. *Āpastamba* improved and expanded on the rules given by Bodhāyana. The general linear equation was solved in the *Āpastamba Śulbasūtra*. His formula gives a remarkably accurate value of '*square root of 2*' correct to five decimal places.

Some Notable Mathematicians of Post-Vedic Period

The tradition of mathematical works has existed continuously from Vedic times, as Vedas were never eclipsed. Non-availability of works of later mathematicians should not be taken as absence. Below we briefly list some outstanding Hindu mathematicians, whose works are prominently available. After briefly describing works of Āryabhaṭṭa I and Varāhamihira, we just list the names of these noted mathematicians.

ĀRYABHAṬṬA - I

Āryabhaṭṭa I is a major name in world history of mathematics. According to internal evidence he was born in CE 476 and died in the year CE 550. His *Āryabhaṭīya* is available. It provides summary of Hindu mathematics known in his time covering topics on arithmetic, algebra, plane trigonometry, continued fractions, quadratic equations, sums of power series and a table of sines. Āryabhaṭṭa gave a systematic treatment of the position of the planets and also gave the radius and period of rotation of the planetary orbits around the Sun. He correctly explains and calculates the Sun and Moon eclipses.

Solution of Quadratic Equations by Āryabhaṭṭa I

Āryabhaṭṭa I has formulated at least two problems whose solution requires solving a quadratic equation. We discuss one of these here, which is on "interest". Before giving Āryabhaṭṭa's general formulation let us consider a numerical example: If eight years interest on the one year's interest together with the interest on $400.00 is $72.00, then what was the interest on $400.00 ?

Let x be the interest on the capital $400.00 for one year.

Now 8 years interest on this interest is $8\ (x/400)x$

According to what is given, we have: $72 = x + 8\ x\ 2/400$

or the quadratic equation

$x^2 + 50x - 3600 = 0$. Giving $x = 40$.

Astronomy

Āryabhaṭṭa I (476-550), in his book *Āryabhaṭīya*, gives the following general formulation of this problem:

मूलफलं सफलं कालमूलगुणमर्ध मूलकृतियुक्तम्।
मूलं मूलार्धोनं कालहृतं स्यात स्वमूल फलम्।।

mūlaphalaṁ saphalaṁ kālamūlaguṇamardha mūlakr̥tiyuktam ।
mūlaṁ mūlārdhonaṁ kālahr̥taṁ syāt svamūla phalam ॥

Given the interest (*phalam*) *x*, on a capital (*mūla*) *c*, together with interest on it is interest (*sa-phalam*) for a certain given period of time *t*, is *s*, then the interest on the capital is given by {[*sct* + (*c*/2)2] - (*c*/2) }/*t*.

VARĀHAMIHIRA

Varāhamihira is practically the most famous astrologer that we have known. Delhi's adjoining area Mehrauli is named after him. He was born in CE 505 and died in CE 587. He worked at Ujjain which had been an important centre for mathematics since around CE 400. The most famous work of Varāhamihira is the *Pañca-Siddhāntika*. It summarises and synthesises five earlier known astronomical treatises. In his calculations, he uses Greek epicycle theory of the motion of the heavenly bodies. Varāhamihira made some important mathematical discoveries. Among these are trigonometric identities, which in present-day notation would mean the following:

$$\sin A = \cos (\tfrac{\pi}{2} - A), \text{Sin}^2 A + \text{Cos}^2 A = 1 \text{ and } 1 - \text{Cos } 2A = 2 \sin^2 A, \text{ etc.}$$

Another important contribution to trigonometry was his sine tables where he improved those given by Āryabhaṭṭa I. He also obtained results on combinations *C*(*n*, *r*), and relations including the one called Pascal triangle formula.

Some Other Major Names in the History of Hindu Mathematics

- Brahmagupta (CE 598-670),
- Bhāskara I (CE 600-680)
- Lalla (CE 720-790)
- Śrīdhara (CE Eighth century)
- Govinda-Swāmī (CE 800-860)

- Mahāvīra (CE 800-870)
- Prathudaka-Swāmī (CE 830-890)
- Śaṅkara (CE 840-900)
- Āryabhaṭṭa II (CE 920-1000)
- Vijayanandī (CE 940-1010)
- Śrīpati (CE 1019-1066)
- Brahma Deva (CE 1060-1130)
- Bhāskara II (CE 1114-1185)
- Mahendra Sūri (CE 1340-1410)
- Nārāyaṇa (CE 1340-1400)
- Mādhava (CE 1350-1425)
- Nīlakaṇṭha (CE 1444-1544)
- Jyeṣṭhadeva (CE 1500-1575)
- Kamalākara (CE 1616-1700)
- Jagannātha (CE 1690-1750)

Conclusion

The birthplace of mathematics, specially that of numbers, geometry and algebra, is India. Origin of numbers, geometry and sequences and series is enshrined in Vedas. Mathematics developed in Vedic tradition in India and from here went to other parts of the world. Mathematics as a philosophical discipline and as a tool for understanding anything was well understood by Indians. To the West through Arabs and to the East by individuals travelling mathematicians and through its great centres of learning. According to Albert Burk, the original proof of Pythagoras theorem was copied by Pythagoras on his visit to India.

References

Sen, Amartya, *The Argumentative Indian*, Penguin Books, 2005.

Rai, Ramniwas (ed), *Āryabhaṭṭa: Āryabhaṭīya*, New Delhi: Indian National Science Academy, 1976.

Atharvaveda, New Delhi: Dayanand Sansthan, 1975.

Cantor, M., *Grako-indische Studien*, Zeit. Fur Math, u., Physik, 22 (1977).

Teresi, Dick, *Lost Discoveries*, Simon & Schuster, 2002.

Hall, W.W.R. *A Short History of Mathematics*, London, 1901.

Menninger, Karl, *Number Words and Number Symbols: A Cultural History of Numbers* (English Translation from original German, by Paul Broneer), Cambridge: MIT Press, 1969.

R̥gveda (Two Parts): New Delhi: Dayanand Sansthan, 1975.

Seidenberg, A. *The Origin of Mathematics*, Archive for History of Exact Sciences, 18 (1978), 301-42.

Yajurveda, New Delhi: Dayanand Sansthan, 1975.

Yajurveda Saṁhitā, ed. Maharshi Devarāṭa, Banaras Hindu University Shodh Prakashan, 1973.

9

An Unpublished Commentary of the Kātyāyana Śulbasūtra

Nabanarayan Bandyopadhyay

In the development of intellectual tradition in ancient India, Sūtra-type literature flourished because of its facility regarding easy memorisation and brevity of expression. We find the texts entitled Śulbasūtras among other texts of the Sūtra period, facilitating the understanding of detailed structure of *vedīs*, *maṇḍapas*, etc., and their method of preparation. Primarily they deal with the rules of demarcation of different types of sacrificial altars, and places for sacred fire, etc. in accordance with the norms and procedures for ritual performances as laid down in the Saṁhitās, Brāhmaṇas and Śrauta-Sūtras, etc. Of the available eight Śulbasūtras, seven belong to the *Kṛṣṇa Yajurveda*, and only one, namely, *Kātyāyana Śulbasūtra* belongs to the *Śukla Yajurveda*. This Śulba Sūtra contains 7 chapters, 101 *sūtras* in prose and 39 verses, concerning necessary rules of geometry. Some editions of the text along with commentaries and translations are available in print. The commentators include Rāma, Karka, Mahīdhara, and Vidyādhara Śarmā. Recently, I have noticed a manuscript of its commentary, written by Gaṅgādhara Pāṭhaka (seventeenth century) who has not been mentioned by any editor/ translator so far. He was the son of Rāmacandra Pāṭhaka and brother of Yājñika Nārāyaṇa, follower of Hariśaṅkara Dīkṣita (teacher of his father); and ancestor of Devabhadra (NCC, 1969, vol. 5). He resided in Stambhatīrtha, i.e. modern Cambey or Khambhat of Gujarat (Wadekar, 2008). There are 6 other works of varied nature to his credit. Interestingly, we can learn the name of Garga before Karka from his introductory verses of benediction. Many charts, diagrams, calculations, etc. relating to mathematics, especially geometry and mensuration, which are conspicuously absent in others' commentaries have been explained in a systematic and scientific way long before the advent of European style of interpretation in this regard. Truly speaking, these Sūtras called Śulbavijñāna (Jha, 2001, 1.9) are our proud possesion of ancient Indian science. Every attempt should be made to collect, preserve, edit and publish these fruits of intellectual tradition. Keeping this view in mind, a comprehensive note on the said commentary is being presented in this paper on the basis of a copy of the manuscript collected by the author.

Introduction

yajñena yajñam ayajanta devās tāni dharmāṇi prathamāny āsan I
te ha nākaṁ mahimānaḥ sacanta yatra pūrve sādhyāḥ santi devāḥ I I

— *Ṛgveda Saṁhitā*, 10.90.16

From the above-mentioned *mantra* it is understood that *yajña* or sacrifice is our *prathama* (*pra-tama*) *dharma*, religion *par excellence*. But how is this to be pursued? Here comes the question of injunctions (*vidhi*s), formuli (*mantra*s), eulogy (*arthavāda*), detailing of norms and procedures, etc. To make the function perfect, Kalpasūtras are written following the prescriptions of the Saṁhitā and the Brāhmaṇas. It is well said: *na tu kalpair vinā kecin mantrabrāhmaṇamātrakāt*. Kalpasūtras occupy an important place so far as Indian religio-scientific tradition is concerned.

We have a long tradition of rites and rituals in India. R.P. Kulkarni (1987) provides us with archeological evidences:

> It is seen from the above-mentioned inscriptional record that the Śrauta sacrifices were being performed, almost without interruption, from Vedic period to Medieval period, throughout India, from Gujarat to Assam and from Magadha in north to Andhra country in the south. Although only one inscription, namely that of Pulakeśin I, could be quoted in which performance of Agnicayana is specifically mentioned, it could be presumed that the kings who had performed important sacrifices like Aśvamedha, Vājapeya, etc., might have performed Soma sacrifice with Agnicayana.
>
> — Preface, p. III

There are evidences of performance of Agnicayana from Archeological excavations:

> Archeological excavations of towns beginning from Palaeolithic period revealed the presence of fireplaces which were presumably meant for fire-sacrifices.
>
> — Introduction, p. IV

Meaning of the Term Śulba

The term *śulba* may be derived from the root *śal* with *auṇādika pratyaya* as per rule *ulvādayaś ca* (4.95) meaning *tāmra* (copper). According to Hemacandra, *śulba* has different meanings like *tāmre yajñakarmaṇy ācāre jalasannidhau*; there is also a root *śulb* (*caurādika*) meaning *māna*. From it, the term conveys the meaning of *rajju* or chord, a root for measurement of altars, fire places, etc. In the development of intellectual tradition in India Sūtra type literature flourished because of its facility regarding easy memorisation and brevity of expression. In the Sūtra period we find the texts of Śulbasūtras among other Sūtra texts since there is prime necessity of understanding the detailed structure of *vedīs*, *maṇḍapas*, etc. and their method of preparation (cf. *śrautayajñārthamaṇḍaparacanātmakaṁ śulbasūtram*). Primarily these Śulbasūtras deal with the rules of demarcation of different types of sacrificial altars and places for sacred fire, etc. in accordance with the norms and procedures for ritual performances as laid down in the Saṁhitās, Brāhmaṇas, and the Śrauta Sūtras, etc. generally speaking, the Śulbasūtras occur in one or more chapters of the Kalpa or Śrautasūtras or are separate. They are of supplementary nature, viz. *pariśiṣṭas*. (Cf. layout of some *vedīs* according to Kātyāyana School for Iṣṭi, Paśu and Soma sacrifices given by Kulkarni, in his book *Layout for Different Sacrifices According to Different Śrauta-Sūtras*, 1997).

Śulbasūtra Texts

Of the available eight Śulbasūtras, seven belong to the *Kṛṣṇa Yajurveda* and one namely, *Kātyāyana Śulbasūtra*, belongs to the *Śukla Yajurveda*. The construction of different altars, *citis*, etc. are to be made by Adhvaryu and his associates. So it is required that Adhvaryu should have geometrical knowledge. That is why Śulbasūtra is not available in other Sūtra texts belonging to the *Ṛgveda*, the *Sāmaveda* and the *Atharvaveda*. (Chaubey, 2004). The *Kātyāyana Śulbasūtra* contains 7 chapters, 101 *sūtras* in prose and 39 verses (41/43 in other's opinion) concerning necessary rules of geometry. The following editions of this Śulbasūtra are available in print:

(1) First two chapters with commentary of Rāma Vājapeyin and English translation were published in the magazine *Pandit* by G.S. Thibaut from Banaras in 1882.

(2) The text with commentaries of Karka and Mahīdhara was published by Chowkhambha Sanskrit Series office and edited by Vedācārya Dogre Anantarama Sastri no. 120.

(3) The text with Śrauta Vṛtti of Vidyādhara Śarmā was published by the Acyuta Granthamala Office of Banaras in 1928, no. 3.

(4) The text with English translation and commentary is included in *The Śulbasūtras of Baudhāyana, Āpastamba, Kātyāyana and Mānava* by S.N. Sen and A.K. Bag, published from Indian National Science Academy, New Delhi, 1983.

(5) The text with English translation and explanatory notes, ed. S.D. Khadilkar, Vaidik Samshodhan Mandal, Poona, 1974.

From the above editions it is evident that the commentators include Rāma Vājapeyin, Karka, Mahīdhara and Vidyādhara Śarmā. Recently I have noticed a manuscript of its commentary

(No. G 5949, The Asiatic Society, Kolkata), written by Gaṅgādhara Pāṭhaka who has not been mentioned by any editor or author excepting Vidyādhara Śarmā. Only two excerpts from Gaṅgādhara's commentary have been incorporated in the footnote of his Vṛtti (pp. 1 & 7). So till now attention has not been given to this commentary by the scholars interested in the study of the Śulbasūtras. That is why I would like to draw the attention of the present learned members of the academic fraternity to take up this commentary for subject of study and research. I am also in serious thought of editing this in coming years as a part of our programme in the School of Vedic Studies, Rabindra Bharati University under the UGC Project on "Studies of Scientific Thoughts and Components in Vedic Literature".

Gaṅgādhara Pāṭhaka and his Works

Now the question may arise — who is this Gaṅgādhara Pāṭhaka? What we can know about his personal history is this: He was the son of Rāmacandra Pāṭhaka and brother of Yājñika Nārāyaṇa, follower of Hariśaṅkara Dīkṣita (teacher of his father); and ancestor of Devabhadra (see *NCC*). He resided in Stambhatīrtha, i.e. modern Cambey or Khambat of Gujarat (Wadekar, 2008). In his introductory verses of benediction he says:

jayaty eko jagatyāṁ śrīhariśaṁkaradīkṣitaḥ |
trayīsāravivektṛśrīrāmacandraguror guruḥ | |
rāmacandraṁ guruṁ natvā gargakarkādiyājñikān |
śrīkātyāyanaśulbasya kurve bhāṣyam idaṁ sphuṭam | |

Here we get a new information about a ritualist named Garga who appeared before Karka. In the colophon it is noted:

rāmacandratanujena gaṅgādharaśarmabhiḥ | |
pitṛcaraṇaiḥ sūtre 'smin kātīye śulbasaṁjñake | |
vivṛtaṁ yat tataḥ śiṣṭaṁ tad iha vivṛtaṁ mayā |
gaṅgādharatanujena rāmakṛṣṇena dhīmatā | |

nāgarasātisaṁjñena kāśipuranivāsinā |
yājñikās tena tuṣyantu prīyatāṁ yajñabhug vibhuḥ |
yad atra sauṣṭhavaṁ kiṁcit tad guror eva kevalaṁ |
yad atrāsauṣṭhavaṁ kiṁcit tan mamaiva guror nahi | |
iti śrīmanmahāyājñikapāṭhakaśrīrāmacandrasūrisunumahāyājñika-pāṭhakaśrīgaṅgādharakṛte kātyāyanaśulbasūtrabhāṣyaṁ
samāptim akārīd iti śivam | |

He has many Vedic works, viz. *Yāgakālaviveka*, *Sarvatomukhapaddhati*, *Yājamānapaddhati*, *Pravaranirṇaya*, etc. to his credit. In *NCC* references to some manuscripts of these texts have been made. He refers to Karka, Mahīdhara, Trikāṇḍamaṇḍanakāra, etc. Many charts, diagrams, calculations, etc. relating to mathematics, specially geometry and mensuration, which are conspicuously absent in others' commentaries have been incorporated and explained fully in a lucid way. I can refer to some of them as available in the manuscript (see diagrams in the MSS — fols. 21b-22a, 33b-34a, 54b-55a noted below).

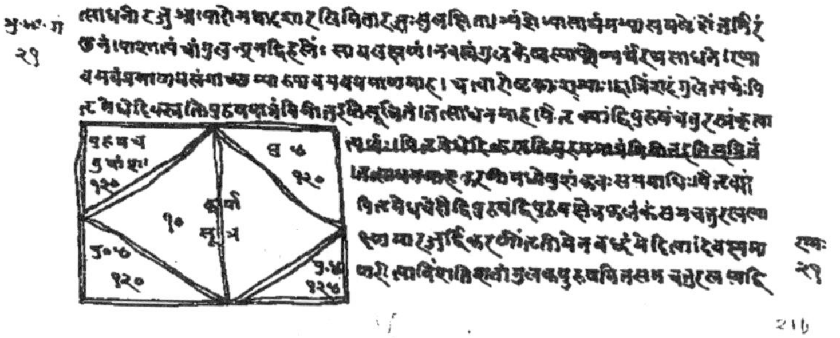

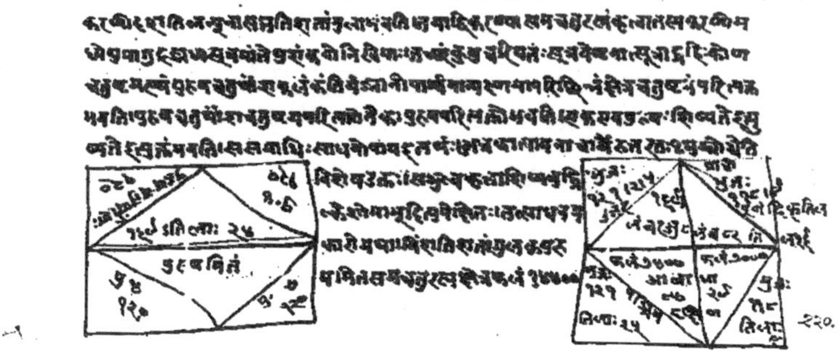

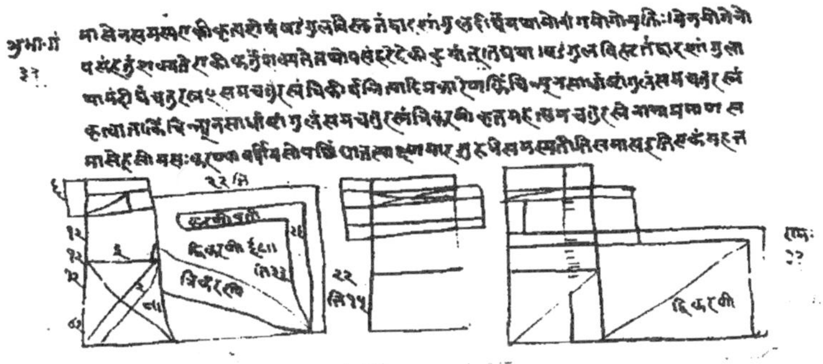

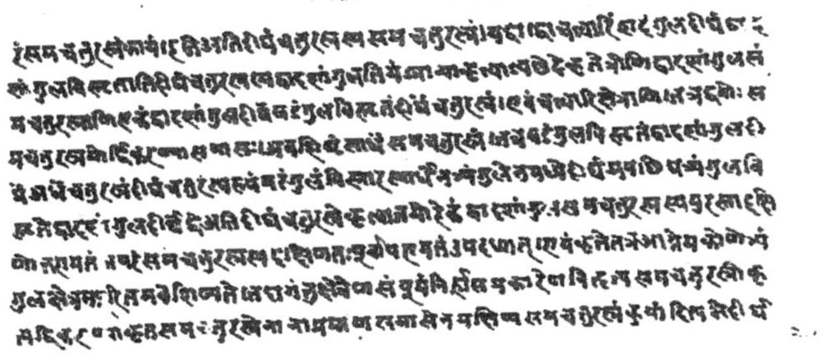

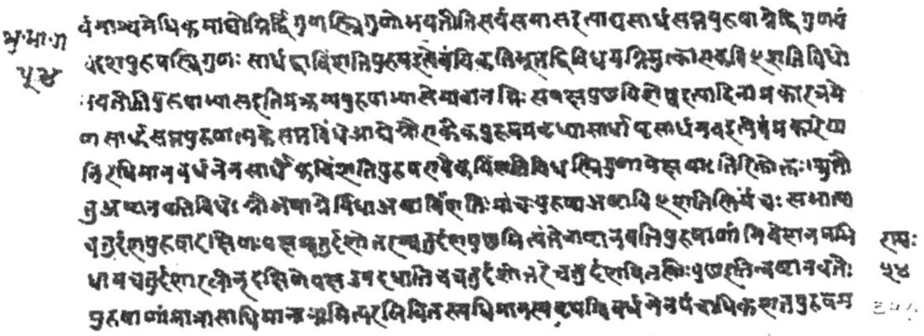

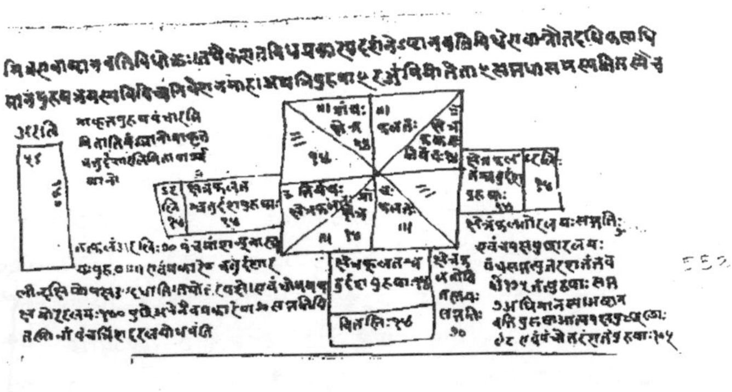

It is astonishing to see his profundity of mathematical scholarship and ability to express his findings in a systematic and scientific way long before the advent of European style of interpretation in respect of ancient Indian geometry.

Regarding availability of diagrams of *citi*s R.P. Kulkarni (1987) in his book (p. III) referred to the diagrams published by Raghuvīra in the *Journal of Vedic Studies* from the manuscripts preserved in the Punjab University Library, Lahore (MSS no. 2021, 4453 and 5846), and Kūrmaciti (MS.35.352 of Bhārat Itihās Saṁśodhaka Maṇḍala, Pune), drawing of *citi* of one square *Puruṣa* (Bundle no. 2/2-3/150 of Prajñā Pāṭhaśālā, Wai). He was also "in possession of xerox drawings which give the arrangement of bricks for each of the five layers of *caturasra śyenaciti* according to the *Kātyāyana Śrauta-Sūtra*" (Ibid., p. 141). It is disheartening to express that R.P. Kulkarni has no knowledge of this manuscript of Gaṅgādhara's commentary full of geometrical figures.

Importance of the Commentaries

We get the full text of commentaries of Karka and Mahīdhara. But nowhere such explanation with the help of charts and diagrams is available. Only the Vr̥tti by Vidyādhara Śarmā of twentieth century contains 10 geometrical figures. On the contrary, Gaṅgādhara's commentary contains more than 50 geometrical figures with interpretation. Herein lies the prime importance of this commentary. I could not go through the commentary of Rāma Vājapeyin; only two chapters have been printed in the *Pandit*. Kundanlal Sharma (1981) in his work *Kalpasūtra* (p. vii) has referred to a paper by S.L. Katre published in the proceedings of the 13th AIOC where it has been mentioned that manuscripts of Rāma's three works are preserved in Scindhia Oriential Institute, Ujjain. I wish the learned scholars to collect copies of these MSS and attempt to edit the texts, namely *Kātyāyana Śulbasūtra Vārttika* containing 515 verses, *Śulbasūtravārttikaṭīkā* (Svopajña) in prose in 6 *Kaṇḍikās*, and *Śulbasūtravr̥tti* in prose in 6

Kaṇḍikās. There are other works of him also entitled *Kuṇḍamaṇḍapalakṣaṇa* and *Nāḍīparīkṣā* on Āyurveda.

Date of Different Commentators

Among the Śulbasūtras, *Baudhāyana Śulbasūtra* is believed to be the earlier. Satyaprakash and Ram Swarup Sharma (1980) opined:

> He must be flourished after the Śatapatha Brāhmaṇa period. He was inspired by the Injunction laid in the Vedic texts. Many of his Sūtras have been quoted by the later authors. . . . We can easily place him some where near 1000 BC to 800 BC." (p. 34).

Thus *Kātyāyana Śrauta-Sūtra* may be dated at least before 800 BCE. According to P.V. Kane, this date, however, comes between 800-400 BCE. Regarding the dates of the commentators of the *Kātyāyana Śrauta-Sūtra*, Karka is earlier (eleventh-twelfth century CE) since he has been referred to by Hemādri (CE 1250) and Trikāṇḍamaṇḍanakāra (CE 1150). Then comes Rāma Vājapeyin, resident of Naimiṣāraṇya. In his introduction to his Vārttika the date of composition CE 1434 has been mentioned. Mahīdhara's date is CE 1589. So Gaṅgādhara is later than Mahīdhara, probably of seventeenth century CE.

Interpretation

Let us take a case of commenting upon the first *sūtra* of the *Kātyāyana Śulbasūtra:* ***rajjusamāsaṁ vakṣyāmaḥ***. Here Karka and Mahīdhara interpret the term *samāsa* only as "additions and subtractions" and *rajju* as "chord". But Gaṅgādhara explains the term *samāsa* as *ekīkaraṇa* and *rajju* as *tatsādhyakṣetra*. He said:

> *tena rajjusamāsaṁ kṣetrasamāsaṁ tulyamānāt tulyapramāṇaṁ kṣetrānām ekīkaraṇaṁ vakṣyāmaḥ* | *samāsagrahaṇam kṣetrapramāṇasamanirhrāsādeḥ śāstrapratipādyasakalaviṣayasyāpy upalakṣaṇārthaṁ tatsarvasyāpy atra vakṣyamāṇatvāt* | |

Since this is a *pratijñāsūtra* so he includes all topics of the concerned Śulbavijñāna here. He further explains elaborately the definition of *rajju* and its characteristics from different sources which are conspicuously absent in others' interpretations. In this way, interpretations of other *sūtra*s have nicely and lucidly been explained with profuse quotations from different texts and commentaries. From the *Kātyāyana Śulbasūtra* along with these commentaries we can gather the following major information:

(a) How to construct *vedīs, citis, śālās,* etc. of a given shape;
(b) Location of the east-west and north-south line and fixing of pegs;
(c) Construction of squares, right-angled triangles, rectangles, trapezium, etc. from different points of calculation;
(d) Measures like *aṅgula, iṣā, akṣa, yuga,* etc.;
(e) Conception of the theorem of Pythagoras;
(f) Definition and value of root 2;
(g) Construction of different *citi*s and increase of the area, measurement, etc.
(h) Importance of the Śulbavijñāna and so on.

Conclusion

Truly speaking, Śulbasūtras called Śulbavijñāna are our proud possession of ancient Indian science. In the *Maitrāyaṇīya-Śulbasūtra*, it has been aptly said: *saurayyavihitā vediḥ sulbavijñānasammitā* (1.9). An expert in Śulbavijñāna has been eulogised by Gopāla Yajvan in his *Śulbarahasyaprakāśana* thus in the following:

> *sūdyogo nyāyavid dhīmān jñātaśrautaḥ suśikṣitaḥ* |
> *śulbānāṁ dhārakaḥ samyak tadarthāt boddhum arhati* | |

Hence every attempt should be made to collect, preserve, edit and publish these fruits of ancient Indian intellectual tradition pertaining specially to the realm of scientific literature by way of discovering hidden truth of the Universe. The sooner we proceed the better:

> *hiraṇmayena pātreṇa satyasyāpihitaṁ mukham* |
> *tat tvaṁ pūṣann apāvṛṇu satyadharmāya dṛṣṭaye* | |
>
> — *Īśa Upaniṣad*, 15

Acknowledgement

I thankfully acknowledge the assistance received from the authorities of The Asiatic Society, Kolkata, regarding supply of xerox copy of the manuscript No. G 5949.

References

Chaubey, B.B., 2004, *Vedāṅga Literature (Auxiliary to the Vedas)*, Hoshiarpur: Kạtyayan Vaidik Sahitya Prakashan.

Jha, Damodar, ed., 2001, *Maitrāyaṇīya-Śulbasūtra* (with the commentary of Śaṅkara), Hoshiarpur: Smt. Priyamvada Jha.

Kulkarni, R.P., 1987, *Layout and Construction of citis according to Baudhāyana-, Mānava-, and Āpastamba-Śulbasūtras*, Research Unit Series No. 10, Poona: Bhandarkar Oriental Research Institute.

———, 1997, *Layout for Different Sacrifices according to Different Śrauta-Sūtras*, Ujjain: Maharshi Sandipani Rashtriya Vedavidya Pratisthan.

New Catalogus Catalogorum (NCC), 1969, vol. 5, University of Madras.

Paṇśīkara, Vāsudeva Śarmā, ed., 2003, *Īśādyaṣṭottaraśatopaniṣadaḥ*, rpt., New Delhi: Rashtriya Sanskrit Sansthan.

Ṛgveda Saṁhitā, 1976, Ajmer: Paropakarini Sabha.

Sharma, Satyaprakash and Ram Swarup, ed., 1980, *Baudhāyana-Śulbasūtram*, 2nd edn., rpt., New Delhi: Mahalakshmi Publishing House.

Sharma, Kundanlal, 1981, *Kalpasūtra*, Vishveshvaranand Indological Series-74, Hoshiarpur: Vishveshvaranand Vedic Research Institute.

Shastri, Haraprasad, *A Descriptive Catalogue of Sanskrit Manuscripts in the Government Collection under the Care of the Asiatic Society of Bengal*, vol. 2 (Vedic Manuscripts), Calcutta: The Asiatic Society.

Wadekar, M.L., 2008, "Contribution of Gaṅgādhara Pāṭhaka — A 17th-Century Scholar from Gujarat", *Aspects of Indological Studies*, Delhi: Bharatiya Kala Prakashan.

Part - IV
Chemical Sciences, Environment and Metallurgy

10

Rasaśālā, the Ancient Indian Pharmacy

A Study on its Construction (Arrangement of Rasaśālā, List of required Apparatuses and Koṣṭhīyantras)

Dillip Kumar Kar

Āyurveda is an accepted system of treating diseases, in human as well as non-humans, prevalent in theory and practice for more than three millennia. From easily available natural resources such as minerals, flora and fauna the medicines were prepared to cure the disorders in men, animals and even trees. All the processes developed for extracting metals like gold, silver, iron, copper, zinc, etc. from their ores, making alloys of these metals with specific properties, using finely powdered and treated metals and metallic preparations in medicine, etc. are essentially chemical processes. Foundation of such practices can be traced back to the pre-historic period as an evidence of finding many metallic articles like axes, daggers, arrow-heads utensils, etc. mainly made of copper, bronze, gold, silver and lead. The practical arts of pottery-making and metallurgical practices like forging, casting smelting, were known to the people of that period.

Introduction

IN the Vedic period we can see positive progress in the field of metallurgical and chemical practices. The references of many metals can be seen in all the four Vedas. The *Atharvaveda* seems to be an important source of such knowledge, in which, many chemical practices are referred for both practical arts and medicinal purposes. Then this practice much developed in the later-Vedic period which reflected in the works like Saṁhitās, Āraṇyakas, Brāhmaṇas and Upaniṣads. The period between 600 BCE to CE 800 can be identified as the golden period as Āyurveda flourished in the first half of this period and many great works like *Caraka Saṁhitā* and *Suśruta Saṁhitā* were produced. In the later half, a firm foundation for Rasaśāstra was laid. Gradually the texts like *Atri Saṁhitā* of Maharṣi Atri; *Dhātusarvasvam* of Aśvinikumārau; *Rasahṛdayatantra* (one of the oldest texts on Rasaśāstra which is written in early eighth century CE) of Govindabhagavatpāda (who was said to be the preceptor of Ādiśaṅkarācārya); *Nāgārjunatantra* (early eighth century CE), *Rasaratnākara, Rasendramaṅgala, Rudrayāmalatantra, Ārogyamañjarī*, etc., of Nāgārjuna (who is famed as the father of Rasaśāstra); *Rasopaniṣad* (CE 1,000) of Somanātha; *Rasendracūḍāmaṇī* (CE 1200) of Somadeva; *Rasaprakāśasudhākara* (CE 1400) of Yaśodhara; *Rasaratnasamuccaya* (CE 1400) of Vāgbhaṭa; *Rasakaumudī* (CE 1500) of Jñānacandra, *Rasasaṁketakalikā* (1600 CE) of Cāmunḍā Kāyastha; *Rasakāmadhenu* (CE 1700) of Śri Cūḍāmaṇi; *Rasacanḍāṁśu* (CE 1700) of Dattārāma Vaidya and the other texts like *Lohārṇavam, Rasajalanidhi, Rasendrapurāṇa, Rasārṇava, Rasārṇavakalpa, Rasamitra*, etc. that deal with different chemical processes came to existence. It was adopted as a traditional practice for fulfilment of livelihood as well as a social service in ancient times but due to the revolutionary changes in the scientific field it forced to them to give up such traditional practices and to search some other new ways of earning.

Āyurveda forms a part of *Atharvaveda* and the Vedic section dealing with the science of medicine is largely lost. However, there are treatises (Saṁhitās), written based on the traditional knowledge system that reflected in the Vedas and the school of Sāṁkhya, the system dealing with this science, authored by Caraka, Suśruta and Vāgbhaṭa being the authorities of this science composed their books for the benefit of world. Besides anatomy, pathology and medicines to treat diseases, the above treatises also deal with the modes of preparation of medicines in the laboratory known as *rasaśālā* (ancient Indian pharmacy).

Location of the Rasaśālā

The *Rasaratnasamuccaya* describes that the place, which is devoid of all types of fears and obstacles, dwelling of Lord Śiva and Goddess Pārvatī, where all the medicinal herbs are easily available, which is beneficial, beautiful to look at, consists of a well and a pond is suitable for constructing the pharmacy (*rasaśālā*). It should also be built surrounded by a beautiful garden with four beautiful gates toward the four directions along with windows, at the northern/north-eastern/eastern part of the site and should be well equipped by all kinds of equipments, apparatuses and chemicals.[1] The *Rasāyanasāra* only describes the view with details.[2]

Arrangement of the Rasaśālā

The treatises like *Rasaratnasamuccaya*, *Rasāyanasāra* and *Rasendrapurāṇa*, etc. describe much on the arrangement of a

---

1. *ātaṅkarahite deśe dharmarājye manorame* |
umāmaheśvaropete samṛddhe nagare śubhe | |
kartavyaṁ sādhanaṁ tatra rasarājasya dhīmatā |
atyantopavane ramye caturdvāropaśobhite | |
tatra śālā prakartavyā suvistīrā manoramā |
samyagvātāyanopetā divyacitrairvivarjitā | | (*ra.ra.sa.* 6.11-13)
rasaśālā prakurvīta sarvabādhāvivarjite |
sarvauṣadhimaye deśe ramye kūpa samanvite | |
yakṣatryakṣasahasrākṣadigvibhāge suśobhite |
nānopakaraṇopetāṁ prakāreṇa suśobhitām | | (*ra.ra.sa.* 7.1,2)
2. *rasāyanasya śālāyāḥ prakāraṁ vacmi pūrvataḥ* |
....
tatra dhanvantarirdeva āyurvedapravartakaḥ | | (*ra. sā.* 2.1-11)

Rasaśālā. As per *Rasaratnasamuccaya* the figure of Rasabhairava should be placed at the eastern (*pūrva*) side, the instruments relating to firework and the operations related to heating, etc. at the south-eastern (*āgneya*) part, grinding works (with the help of mortar and pestle of various materials) at the southern (*yāmya*) part, the process using sharp instruments at the south-western (*nairṛtya*) part, the work relating cleaning, washing, etc. at the western (*vāruṇa*) region, the work of drying at the north-western (*vāyu*) corner, the work related to cutting, beating, hitting, etc. at the northern (*uttara*) region and the north-eastern (*īśa*) part of the workshop should be left for the storage of prepared medicines as well as raw minerals, herbs as well as herbal products and chemicals.[3]

On the establishment of God Rasabhairava the *Rasaratnasamuccaya* says that it should be established at the eastern part (most probably at the entry point and back to which the entire pharmacy is situated) whereas the *Rasāyanasāra* instructs it to establish at the centre of the pharmacy (most probably facing towards the east).[4] In the case as instructed by *Rasaratnasamuccaya* the first vision of a worker falls on the god at the gate before entering into the pharmacy whereas in the case of *Rasāyanasāra* the workers meet to the god just after entering into the pharmacy. However, in both the cases it is

---

3. *śālāyāḥ pūrvadigbhāge sthāpayedrasabhairavam* |
vahnikarmāṇi cāgneye yāmye pāṣāṇakarma ca | |
nairṛtye śastrakarmāṇi vāruṇe kṣālanādikam |
śoṣaṇaṁ vāyukoṇe ca vedhakarmottare tathā | |
sthāpanaṁ siddhavastūnāṁ prakuryādīśakoṇake |
padārthasaṅgrahaḥ kāryo rasasādhanahetuka | | (*ra.ra.sa.* 7.3-5)

4. *catuśśālasya mdhye tu devatāyatanambhavet* |
tatra dhanvantvarijdeva āyurvedapravartakaḥ | |
bhairavapramukhāścānye śaṅkaraśca jagadguruḥ |
rasāyanasyādyācāryaḥ prayatamānase | | (*ra. sā.* 2.11-12)

observed that the figure of the God Rasabhairava is instructed to be established in such a place to which the vision of a worker will go automatically so that he can offer his prayer to the god to get blessings for better output before he starts his chemical operations inside the pharmacy.

In Indian mythology the rain god Indra is said to be the lord of the eastern direction (*dikpāla*) as a result the eastern part of this sacred land yields good crop. The establishment of the God Rasabhairava at the eastern part of the pharmacy (*rasaśālā*) is as a means of more productivity. *Vahni*, the fire god, is the lord (*dikpāla*) of south-eastern corner (*āgneya koṇa*), which shows that the power of the fire god in this region is more effective. Similarly, Yama is the *dikpāla* of south direction (*dakṣiṇadik*). Here Yama is the god of death and Yamaloka (the land of Yama) is the place where the *jīvātmā* gets punishment after his death. Therefore the work relating to rubbing (*gharṣaṇam*), grinding (*peṣaṇam*) by stones, etc. is normal in this land and as a result it is believed that such stone-related works will be more effective in this south direction. In the same way the lord (*adhipati*) of south-western corner (*nairr̥tya koṇa*) is Nairr̥ta, a demon. It is observed in all Purāṇas that the demons are always inclined towards war and they try to have new and more powerful weapons, possibly, due to the reason to please him and get his better blessings in the form of a better output, this south-western direction is selected for the works relating to weapons, i.e. cutting, holing, forging, quenching, etc. The god of waters, Varuṇa, is the *adhipati* of western direction (*paścimadik*). The presence of Varuṇa in the west shows the sufficiency of water in this direction and therefore it has been chosen to do the works related to water such as washing, cleaning, etc. in this direction. Vāyu is the *adhipati* of north-western direction *vāyavya koṇa* that shows the presence of *vāyu* in this corner who helps to dry and in the winnowing work. The lord of north direction (*uttaradik*) is Kubera. Bila[5] is one of the names of Kubera which is related to the actions like cutting, holing, making into many pieces, etc. and therefore most probably, to get blessings from him in the form of better output the works related to cutting, holing, making many pieces, etc. (probably to the raw drugs) is instructed to do in the north region of the pharmacy (*rasaśālā*). Īśa is said as the god of north-eastern corner (*aiśānya koṇa*). Īśa is mentioned as a universal god (Vaiśvadeva) who is very much fond of acquiring things. Possibly to please him, this direction is left for the storage of all types of raw herbs, chemicals and minerals as well as the prepared medicines.

On the scientific point of view if the above picture will be analysed, all processes which are mentioned to perform inside

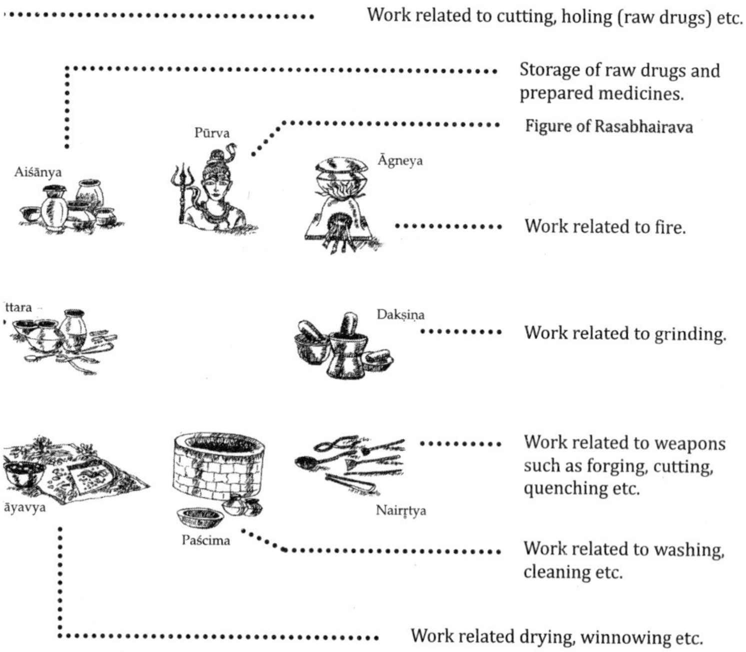

fig. 1: (Arrangement of Rasaśālās per Rasaratnasamuccaya)

5. *a. ko.* – 1.1.160

a pharmacy (*rasaśālā*) shall be found interlinked in an anti-clockwise order. For example, just after entering (entry from the east direction, i.e. *pūrva*) into the room the right side (north-east — *aiśānya*) region is specified for the storage of raw drugs and prepared medicines. The storekeeper supplies raw drugs from the store for preliminary processing, i.e. for cutting and holing, etc. (make into pieces) to *uttara* region, then the cutter (person who cuts) sends the materials to *vāyavya* region after primary processing for drying and winnowing. Then for washing by water/cleaning it goes to *vāruṇa/paścima* region then to make those smaller by sharp weapons to *nairṛtya* region and after that he sends those to *yāmya/dakṣiṇa* direction to grind for making them fine powders or to mix all the herbs, minerals and chemicals proportionately. After this processing the chemist of *āgneya* region receives them from the worker of the *yāmya/dakṣiṇa* direction and finally prepares different medicines by using different stoves/furnaces. As soon as the preparation completed these are sent away to serve the patients through the east gate or to store in the storage at the north-east corner.

The *vāyavya koṇa* is situated just to the opposite of *āgneya koṇa* and air always blows from *vāyavya koṇa* towards the *āgneya*

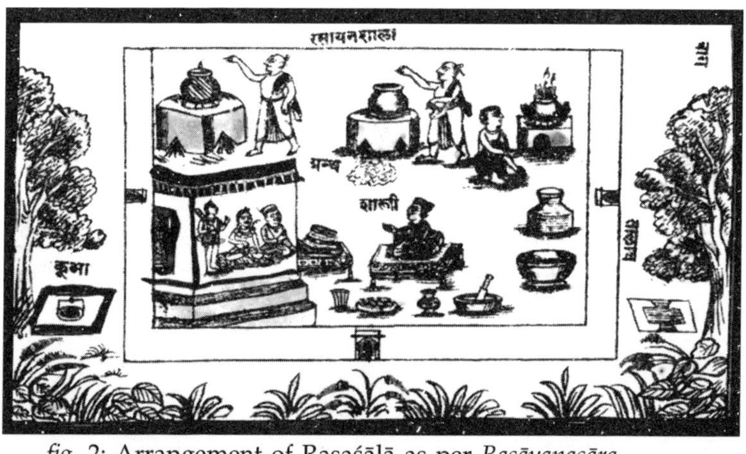

fig. 2: Arrangement of Rasaśālā as per *Rasāyanasāra*

koṇa, as a result the inner part of *rasaśālā* always gets free from heat and smoke as well as the furnaces also burn with high flames which leads the workers to work more effectively without any disturbance. Similarly the air also helps the workers for drying and winnowing different things more effectively at *vāyavya koṇa.*

Appointment of Different Employees in the *Rasaśālā* and their Required Qualifications

For the proper functioning of this workshop, it is instructed in almost all Āyurvedic texts to appoint different employees like chemists, physicians, drug-collectors and attendants, etc. in it.

The persons who are truthful, free from temptations, devotees of *deva*s and brāhmaṇas, self-controlled and used to live upon proper diet and regimen, are to be engaged for performing chemical operations. Such herbalists who are not deceitful and are well-versed in the knowledge of the drugs and plants in different languages of many countries should be employed.[6]

The physician who is to be appointed in the said pharmacy, should have studied the Rasaśāstra well, should have proper access to materia-medica (Nighaṇṭuśāstra) and also he should be well versed in the languages of all the (nearby) countries.[4] Hence, he should be pious, truthful, scholarly, devotee of Lord Śiva and Viṣṇu, kind-hearted and should have lotus sign on his palm.[7]

The physician who is having in his palm one or more signs such as flag, sacred vessel, lotus, fish and bow; or who is having

6. *rasacintakā vaidyā nighaṇṭujñāśca vārtikāḥ* |
sarvadeśajabhāṣajñāḥ saṅgrāhyāste 'pi sādhakaiḥ | |
dharmiṣṭhaḥ satyavān vidvān śivakeśavapūjakaḥ |
sadayaḥ padmahastaśca saṁyojyo rasavaidyake | | (ra.ra.sa. 7. 24 & 27)

7. *dharmiṣṭhaḥ satyavān vidvān śivakeśavapūjakaḥ* |
sadayaḥ padmahastaśca saṁyojyo rasavaidyake | | (*ra.ra.sa.* 7.27)

lines up to base of the ring finger, is called *amṛta-hasta-puruṣa*.[8] In Āyurveda such physicians are considered as the best ones.

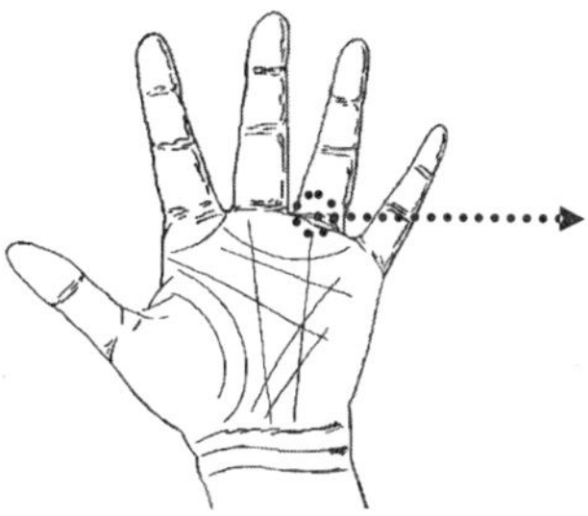

The physician who has any one of the following five symbols in his hands is considered as the best one.

1. Patākā – Flag, 2. Kumbha – A pot with narrow neck, 3. Pāthoja – Lotus, 4. Matsya – Fish, 5. Cāpa – Bow

fig. 3: Symbols of a Good Physician

That physician who has no country of his own, who is cruel and greedy in nature, without having education from a teacher and who has black-coloured lines on the palm, is called *dagdha-hasta-puruṣa*.[9] In Āyurveda the appointment of such physicians in a pharmacy is strictly restricted.

The person who has the knowledge of the advices (written in our sacred texts or told by our forefathers) which are known as *mantra*s (as the root *mantra* means *upadeśa* or advice) by which he can control or break the natural disasters (*bhūta-nigraha-mantrajñaḥ, bhūta* — the five great elements, i.e. earth, wind, water, fire and space, *nigraha* – to control, to break, etc. *mantrajñaḥ* — who knows the advices), should be appointed as the collector of different drugs.[10]

The person who is industrious, having good conduct and be neat and clean, strong and courageous, should be appointed as the attendant in the said workshop.[11]

---

8. *patākākumbhapāthojamatsyacāpāṅkapāṇikaḥ* |
anāmādhaḥstharekhāṅkaḥ sa syādamṛtahastavān | | (*ra.ra.sa.* 7.28)

9 *adeśikaḥ kṛpāmukto lubdho guguvivarjitaḥ* |
kṛṣṇarekhākaro vaidyo dagdhahasta sa ucyate | | (*ra.ra.sa.* 7.29)

10. *bhūtanigrahamantrajñāste yojyā nidhisādhane* | | (*ra.ra.sa.* 7.30)

11. *sodyamāḥ śucayaḥ śurā baliṣṭhāḥ paricārikāḥ* | | (ra.ra.sa. 7.26)

A person who has the knowledge of different names of medicinal herbs, possesses good conduct, who is without fraud and who has the knowledge of many things and different languages, can be appointed for the collection of raw drugs.[12]

For some specific type of chemical operations the chemist requires the organics such as bloods, fleshes, bones, marrows, etc. of some specific animals or birds and for which he needs a person who will kill such animals for him whenever he requires. The work of killing animals for the benefit of this earth is called sacrifice (*bali-karma*). A person who is strong, truthful, having red eyes and body with black complexion, and who has the knowledge to make fear among the other creatures (*bhūtatrāsana, bhūta* — creatures, *trāsana* — to generate fear), should be appointed for this *bali-karma*.[13]

A person who is rich, has sacrificing mentality, possesses all types of instruments and who follows the advice of teachers and preceptors is good to appoint for Dhātuveda (metallurgy or alchemy).[14]

List of Required Apparatuses

The *rasaśālā* should be well equipped by different basic apparatuses.[15] As per *Rasaratnasamuccaya* the basic apparatuses are as follows:

---

12. *tattadauṣadhanāmajñāḥ śucayo vañcanojjhitāḥ* |
nānāviṣayabhāṣājñāste matā bheṣajāhr̥tau | | (*ra.ra.sa.* 7.34)
13. *baliṣṭhāḥ satyavantaśca raktākṣāḥ kr̥ṣṇavigrahāḥ* .
bhūtatrāsanavidyāśca te yojyā balisādhane | | (*ra.ra.sa.* 7.31)
14. *dhanavanto vadānyāśca sarvopaskarasaṁyutāḥ* |
guruvākyaratā nityaṁ dhātuvādeṣu te śubhāḥ | | (*ra.ra.sa.* 7.33)
15. *sattvapātanakoṣṭhī ca surākoṣṭhī suśobhanām* |
bhūmikoṣṭhī calatkoṣṭhī jaladroṇyo 'pyanekaśaḥ |
bhastrikāyugalaṁ tadvannālike vaṁśalohayoḥ | |
svarṇāyoghoṣaśulbāśmakuṇḍyaścarmakr̥tāṁ tathā |
karaṇāni vicitrāṇi dravyāṇyapi samāharet | |

→

1. *koṣṭhī* (stove/furnace)
   i. *sattvapātanakoṣṭhī*
   ii. *surākoṣṭhī*
   iii. *pātālakoṣṭhī/bhūmikoṣṭhī*
   iv. *calatkoṣṭhī* (portable furnace/stove)
   v. *aṅgārakoṣṭhī*
   vi. *gārakoṣṭhī*
   vii. *tiryakpātanakoṣṭhī/mūṣākoṣṭhī.*
2. *jaladroṇī* (water containers of different sizes)
3. *bhastrikā* (two leather bellows attached with bamboo and metal pipes)
4. *kuṇḍa* (vessels made of gold, iron, copper, bronze, leather and stone)
5. *kaṇḍāṇī* (small pounder)
6. *peṣaṇī* (stone grinder)
7. *droṇī* (big cauldron)
8. *taptakhalva* and *mardaka* (a pair of metal mortar and pestle which can be heated on fire)
9. *cālanī* (sieve)
10. *kaṭatrāṇī* (utensils made of stone or metal)

→ *kaṇḍāṇī peṣaṇī khalvā droṇīrūpāśca varttulāḥ* |
āyasāstaptakhalvāśca mardakāśca tathāvidhāḥ | |
sūkṣmacchidrasahasrāḍhyā dravyagālanahetave |
cālanī ca kaṭatrāṇi śalākā hi ca kuṇḍalī | |
mūṣāmṛttuṣakārpāsavanopalakapiṣṭakam |
trividhaṁ bheṣajaṁ dhātujīvamūlamayaṁ tathā | |
śikhitrā govarañcaiva śarkarā ca sitopalā |
kācāyomṛdvarāṭānāṁ kūpikā caṣakāṇi ca | |
śūrpādi veṇupātrāṇi kṣudrakṣiprāśca śaṅkhikāḥ |
kṣuraprāśca tathā pākyau yaccānyattatra yujyate |
pālikākarṇikā caiva śākacchedanaśastrakāḥ | | (*ra.ra.sa.* 7.6-13)

11. *śalākā* (a type of knife)
12. *kuṇḍalī* (circular disks to keep the vessels on)
13. *mūṣā* (crucible)
14. *mr̥ttikā* (soil)
15. *tuṣa* (paddy husk)
16. *kārpāsa* (cotton),
17. *vanopala* (cow-dung cakes)
18. *piṣṭaka* (flour)
19. *trividhaṁ bheṣajam* (three types of drugs)
    i. *dhātujam* (mineral products)
    ii. *jīvajam* (animal products)
    iii. *mūlamayam* (herbal products)
20. *śikhitrā* (charcoal)
21. *govaram* (dried cow-dung powder)
22. *śarkarā* (cane-sugar/sugar)
23. *sitopala* (candy sugar)
24. *kāca-varāṭa*, and *mr̥d-varāṭa* (bottles made of glass and mud)
25. *kūpikā* (small pot)
26. *caṣaka* (drinking glass)
27. *sūrpa* (winnowing baskets made up of bamboo)
28. *kṣudraśiprā* (small pearl shells)
29. *kṣurapra* (sickles)
30. *pālikā* (a sharp edged blade)
31. *karṇikā* (a small knife, shaped like an earring)
32. *śākacchedanaśastraka* (knives to cut the herbs)

All these equipments, starting from the cleaning of the pharmacy up to the *rasapāka* (final process regarding mercurial operations), should be made available and stored in the pharmacy. These too are worshipped; chanting by *śrī-rasāṅkuśa-mantra*[16] failing which the God Bhairava may destroy the power of these equipments. However, the instruction to worship the equipments is nothing but to create a feeling on the sacredness of these equipments in the minds of the workers and as result they will take intensive care of those instruments so that no harm will come on them and the chemical operations will run very smoothly. Hence, the same text also instructs to chant *aghora-mantra*[17] till the completion of each chemical operation for getting good and effective medicines, which indicates towards the creation of attentiveness in the minds of the workers towards their own duty.

In order to extract the metallic portions (*sattva*) from their ores as well as to purify them, furnaces/stoves *(koṣṭhī-yantras*)[18] of different sizes and shapes are required. Such furnaces/stoves are described hereafter.

SATTVAPĀTANAKOṢṬHĪ

The *Rasendrapurāṇa* says about three types of *sattvapātana-koṣṭhī-yantras*. Those are as follows:

TYPE I

This furnace is shaped like a bucket. It is slanted downwards by 8 *aṅgulas* in each side (*aṣṭasaṅkhyāṅgulaṁ tiryak*) and made of iron or mud; placed upon a smooth-surfaced ground; plastered all over with mud; and furnished with two circular air passages — the lower one being just as to fit the end of the

16. *śālāsammārjanādyaṁ hi rasapākāntakarma yat |*
tatropayogi yaccānyattatsarvaṁ paravidyayā | |
śrīrasāṅkuśayā sarvaḥ mantrayitvā samarcayet | |
anyathā tadgataṁ tejaḥ parigṛhṇanti bhairavāḥ | | (*ra.ra.sa.* 7.22-23)
17. *rasapākāvasānāṁ hi sadāghoraṁ ca jāpayet | |* (*ra.ra.sa.* 7.25)
18. *satvānāṁ pātanārthāya pātitānāṁ viśuddhaye |*
koṣṭhikā vividhākārāstāsāṁ lakṣaṇamucyate | | (*ra.ra.sa.* 10.32)

tube of a bellow. Fill it with sufficient quantities of charcoal and blow the fire by means of a tube, made of bone of an animal fixed with a bellow. It is used especially for extracting essences of mica.[19]

"Aṣṭasaṅkhyāṅgulaṁ tiryak" it means the diameter of the lower part is 8 aṅgulas i.e. 6 inches/15.68 cms. less from each side. So the length of this lower diameter is 48-(8+8) = 32 aṅgulas i.e. 24 inches/62.72 cms.

"Hastapramāṇaṁ dīrghāram" it means the radius (distance from point A to B) of its upper part is one hand i.e. 24 aṅgulas = 18 inches/47.04 cms. Therefore length of its diameter (distance from point A to C) will be 1 + 1 = 2 hands i.e. 48 aṅgulas = 36 inches/94.08 cms.

The distance from point A to D and C to H = 1 hand i.e. 24 aṅgulas = 18 inches/47.04 cms.

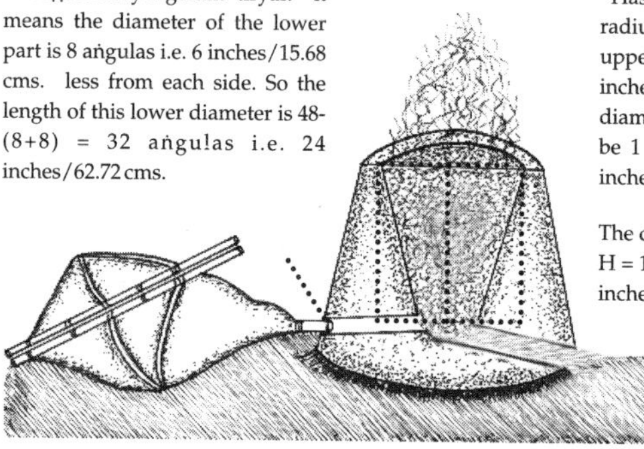

fig. 4: Sattvapātanakoṣṭhī, Type QI

TYPE II

Prepare a *khalva*-shaped oven, upon which another *khalva* (a boat-shaped mortar having an egg-shaped pit in its middle portion, which is specially used to grind drugs) made up iron is to be placed. The oven is to be filled with charcoal, set on fire, which is to be blown by means of a bellow, provided at one side of it. Mercurial cake, rubbed with *kṣāra*s, *amla*s, etc. gets quickly smelted in it. In case, the mercurial cakes are heated in a *khalva* (upper one) made up *kāntāyasa*[20] they become very effective.

19. *hastapramāṇaṁ dīrghāramaṣṭasaṅkhyāṅgulaṁ tiryak* |
samabhūbhāge ghaṭitaṁ vṛttaṁ mṛtkarmasampannam | |
vātāyanadvīvṛttaṁ bhastrīmukhacūyadhobhāge |
pradhametpaśuvasānālairbhastrābhirvā 'bhrasattvārtham | | (*ra. pu.* 2.146-47)

20. *kṛtvā khalvākṛtiṁ cūllīmaṅgāraiḥ paripūritām* |
tasyāṁ nivesitaṁ khalvaṁ pārśve bhastrikayā dhamet | |
rasena marditā piṣṭaḥ kṣārairamlaiśca saṁyutaḥ |
pradārayatyativegena sveditā nātra saṁśayaḥ | |
kṛtaḥ kāntāyaso 'yaṁ bhavetkoṭiguṇo rasaḥ | | (*ra. pu.* 2.148-50)

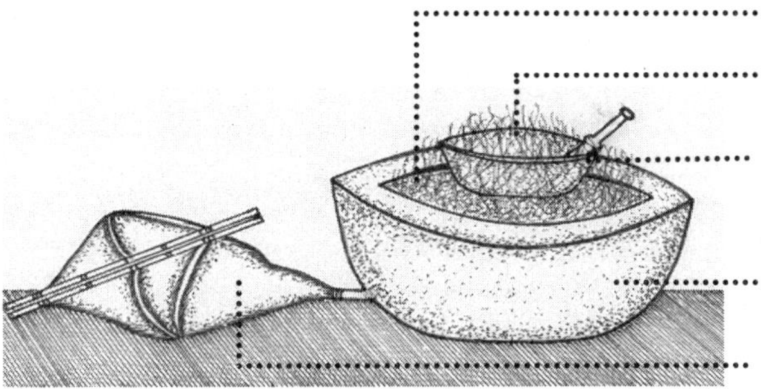

fig. 5: Sattvapātanakoṣṭhī, Type II

TYPE III

It is a *koṣṭhī,* of 16 *aṅgula*s high and 12 *aṅgula*s in length and width, which is generally used for the purpose of extracting essences of different metals. Fill the furnace with the charcoal prepared from the wood of bamboo (*vaṁśa*), *khadira, madhuka* and *badarī*. Then fire the charcoal and blow air into it through the internal passage by a bellow.[21] The *Atri Saṁhitā*[22] and *Rasaratnasamuccaya*[23] also view the same about this furnace.

The *Rasaratnasamuccaya* mentions about some more furnaces. Those are as follows:

---

21. *śoḍaṣāṅgulavistīrṇā hastamātrāyatā samā* |
dhātu sattvanipātārthaṁ koṣṭhikā parikīrttitā | |
vaṁśakhādiramādhūkabadarīdārusambhavaiḥ |
paripūrṇā dṛḍhāṅgārairadhovātena koṣṭhake | |
bhastrayārantaḥmārgeṇa jvālayecca hutāśanam | | (*ra. pu.* 2.151-53)

22. *śoḍaṣāṅgulavistīrṇā hastamātrāyatam samam* | |
dhātu sattvanipātārthaṁ koṣṭhikam parikīrttitam |
paripūrṇā dṛḍhāṅgārairadhovātena koṣṭhake | |
mātrayā jvālamārgeṇa jvālayecca hutāśanam | | (*a. saṁ*)

23. *śoḍaṣāṅgulavistīrṇā hastamātrāyataṁ samam* |
dhātu sattvanipātārthaṁ koṣṭhiyantramiti smṛtam | | (*ra.ra.sa.* 9.43)

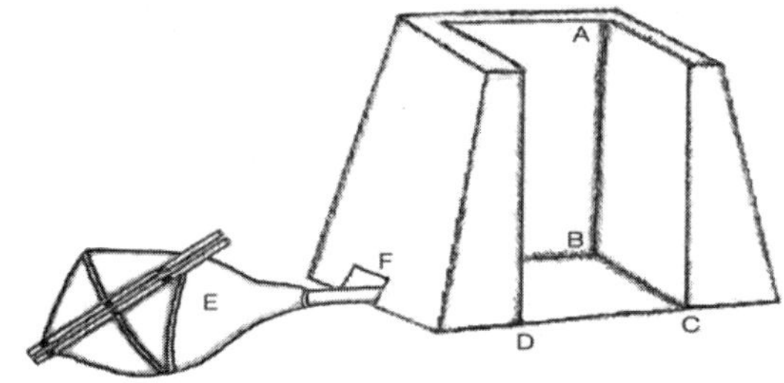

- The distance between A and B = 1 hasta (24 aṅgulas) i.e. 18 inches/47.04 cms.
- The distance between the points B to B and C to D = 16 aṅgula i.e. 12 inches/31.36 cms.
- E - Bhastrīkā
- F - Gate

fig. 6: Sattvapātanakoṣṭhī, Type III

AṄGĀRAKOṢṬHĪ

Construct an earthen cabin on the earth measuring one *rājahasta* (about 30 *aṅgula*s) in height, ½ *rājahasta* (about 15 *aṅgula*s) in width and ½ *rājahasta*s in width (about 15 *aṅgula*s). By leaving 1 *vitasti* (12 *aṅgula*s) space from the bottom of one of its walls, an opening is made measuring ½ *vitasti* (6 *aṅgula*s) in length and width. Necessary arrangements should be done to blow air into the furnace through this passage by means of a bellow. In the roof of the construction makes another passage measuring one *pradeśa* (the distance between the tips of the little finger and thumb finger when the palm is expanded, i.e. about 7.5") in length and width to upwards and close remaining open space by mud/bricks and scale it properly by mud plaster.

E – Gate to blow air, O – Upper gate

I & J - Bhastrikās

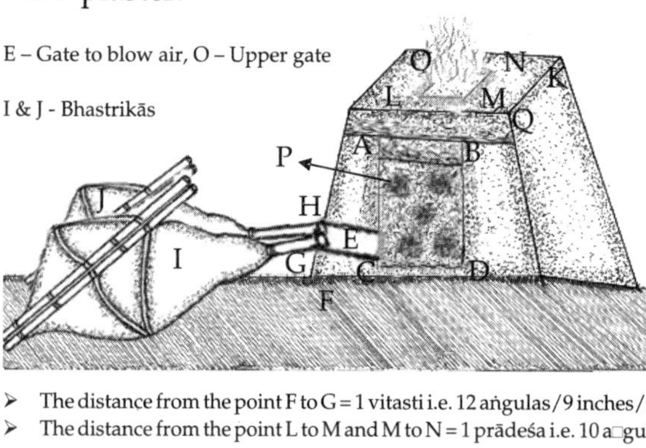

- The distance from the point A to C = 1 rājahasta i.e. 30 aṅgulas/22.5 inches/58.8 cms.
- The distance from the points C to D and Q to R = 1/2 rājahasta i.e. 15 aṅgulas/ 11.25 inches/ 29.4 cms.
- The distance from the point F to G = 1 vitasti i.e. 12 aṅgulas/9 inches/23.52 cms.
- The distance from the point L to M and M to N = 1 prādeśa i.e. 10 a□gulas/7.5 inches/19.6 cms.

fig. 7: Aṅgārakoṣṭhī

The burning charcoal is put into the furnace (*koṣṭhī*) from the top opening and bellowed air with the help of two bellows. Then the mineral drugs which are made into the spherical balls (for *sattvapātana*), also put inside and bellowed intensively. Generally such 5 balls are placed at a time. This is called *aṅgārakoṣṭhī*, which is useful in the *sattvapātana* of harder minerals.[24]

PĀTĀLAKOṢṬHĪ/BHŪMIKOṢṬHĪ

A round pit measuring one *vitasti* (12") deep and equally wide is made inside the earth. Another small pit 4 *aṅgula*s in length and 4 *aṅgula*s in width is dug inside the above bigger pit in the centre. A twisted pipe is set so that the one goes inside the smaller central pit and one outside the bigger pit. This end opens at the surface of the earth and can be connected to the centre. Then, a round earthen disk or wheel (*cakra*) with five holes is placed upon the central pit and one the outer pit is filled with coal. (After placing the *sattvapātana* materials) the coal is ignited and air is pumped with the help of bellows. It is called *pātālakoṣṭhī* and used for *sattvapātana* of softer

24. *rājahastasamutsedhātadardhāyāmavistarā* |
caturasrā ca kuḍyena veṣṭitā mṛṇmayena ca | |
ekabhittau caredvāraṁ vitastyābhogasaṁyutam |
dvāraṁ sārdhavitastyā ca sammitaṁ sudṛḍhaṁ śubham | |
dehalyadho vidhātavyaṁ dhamanāya yathocitam |
pradeśapramitā bhittiruttaraṅgasya corddhvataḥ | |
dvāraṁ copari karttavyaṁ prādeśapramitaṁ khalu |
tataścesṭikayā ruddhā dvārasandhivilipya ca | |
śikhitraistā samāpūrya dhamedbhastrādvayena ca |
śikhitrān dhamanadravyamūrddhvadvāreṇa nikṣipet | |
sattvapātanagolāṁśca pañca pañca punaḥ punaḥ |
bhavedaṅgārakoṣṭhī sā kharāṇāṁ sattvapātanī | | (*ra.ra.sa.* 10.33-38)

minerals. It is described by the scholar Nandī and is convenient for extracting the *sattva* by very less blowing (heating).[25]

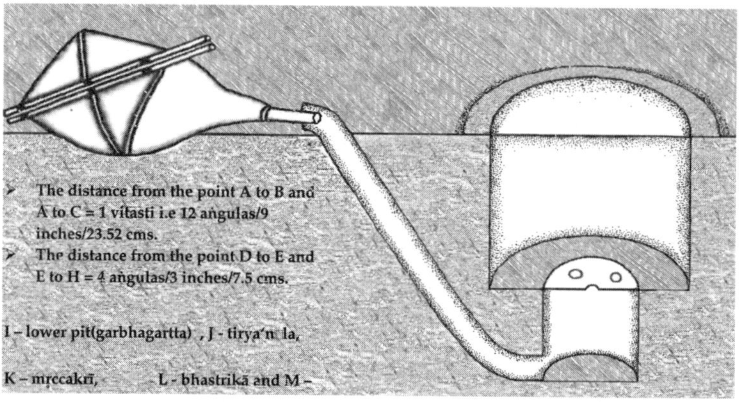

fig. 8: Pātālakoṣṭhī/Bhūmikoṣṭhī

GĀRAKOṢṬHĪ

A cylindrical construction is made measuring 12 *aṅgula*s in height and one *prādeśa* in width. Then from the bottom of this construction, attaching to the inner wall another cylindrical construction is made measuring 4 *aṅgula*s in height and width. The top of the inner construction will appear like a ring. On this ring a disk (*cakra*) having several holes, made of metal or mud is placed. In the bottom of the construction a gate is also made to blow air. Charcoal is filled in the upper pit upon the wheel and (after placing the crucible (*mūṣā*) filled by powdered metals and chemical, covered by charcoals) fire is given and bellowed with the help of

25. *dṛḍhabhūmau caredgarttaṁ vitastyāsammitaṁ śubham* |
varttulañcātha tanmadhye garttamanyaṁ prakalpayet | |
caturaṅgulavistāranimnatvena samanvitam |
garttāddharaṇiparyantaṁ tiryaṅnālasamanvitam | |
kiñcitsamunnataṁ bāhyaṁ garttābhimukhanimnagam |
mṛccakrī pañcarandhrāḍhyā garbhagarttodare kṣipet | |
āpūrya kokilaiḥ koṣṭhī pradhamedekabhastrayā |
pātālakoṣṭhikā hyeṣā mṛdūnāṁ sattvapātanī |
dhmānasādhyapadārthānāṁ nandinā parikīrttitā | | (*ra.ra.sa.* 10.39-42)

vaṅkanāla. This *gārakoṣṭhī* is used for separating the metallic to eradicate the blemishes from the metals.[26]

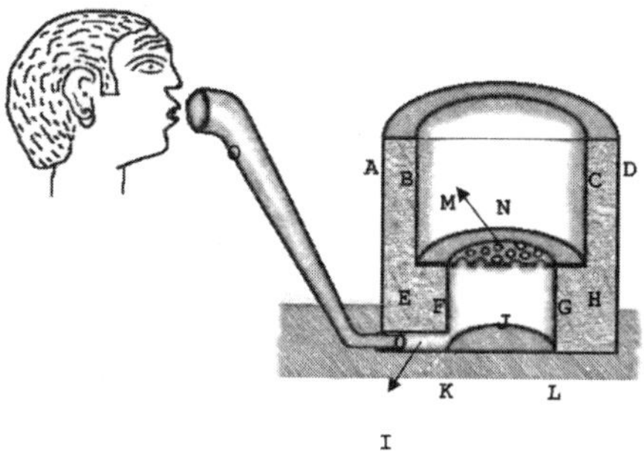

- "dvādaśāṅgulanimnā" The distance between the lines BC and KL = 12 aṅgula i.e. 16 inches/23.52 cms.
- "prādeśapramitā tathā" the distance between B and C = 1 prādeśa i.e. approximately 10 aṅgulas/7.5 inches/19.6 cms.
- "caturaṅgulataścordhve" the distance between the lines FG and KL = 4 aṅgulas/3 inches/7.84 cms.

N – Upper pit, EFGH – valaya, M – Disk (bhūrichidravatī cakrī), I – gate, O – vaṅkanāl and J – lower pit

fig. 9: Gārakoṣṭhī

TIRYAKPĀTANAKOṢṬHĪ

A slanted cylindrical furnace is constructed measuring 12 *aṅgula*s in height with same length of the diameter at top and with the length of 4 *aṅgula*s of diameter at the bottom. A passage is left at its bottom with which a bent pipe is to be connected to blow air into it. This is called *tiryakpātanakoṣṭhī* (or *mūṣākoṣṭhī*) and it is used for the purification of soft metals.[27]

In the first part of this paper, attempt has been made to simplify the technicalities of constructing an ancient Indian pharmacy. Except the first two pictures of pharmacy (*rasaśālā*) (the first picture is taken from the book *Science and Technology*

---

26. *dvādaśāṅgulanimnā yā pradeśapramitā tathā* |
caturaṅgulataścordhve valayena samanvitā | |
bhūricchidravatī cakrī valayopari nikṣipet |
śikhitrāṁstatra nikṣipya pradhamedvaṅkanālataḥ | |
gārakoṣṭhīyamākhyātā sṛṣṭalohavināśinī | | (*ra.ra.sa.* 10.43-44)
27. *koṣṭhī siddharasādīnāṁ vidhānāya vidhāyate* |
dvādaśāṅgulakotsedhā sā budhne caturaṅgulā | |
tiryakpradhamanāsyā yā mṛdudravyaviśodhinī | | (*ra.ra.sa.* 10.46)

in Ancient India through the Ages which is based on my previous research work, published by the Academy of Sanskrit Research, Melkote and the second picture is scanned from the book *Rasāyanasāra* written and published by Paṇḍita Śyāmasundarācārya Vaiśya (Rasāyanaśāstrī), Rasāyanaśālā, Kāśī, 1915.) all the other line-drawings are drawn as per the measurements given in the referred *ślokas*. Hope, these will give some new directions to the modern scholars for its further study. I personally request for the comments on these designs so that the errors can be rectified properly. The measurements of which furnaces are found complete from all sides are lined out here and which are found incomplete are kept for further study. In the second part of this paper attempt will be taken to collect all the information on the instruments and apparatuses, etc. those are required for an ancient Indian pharmacy.

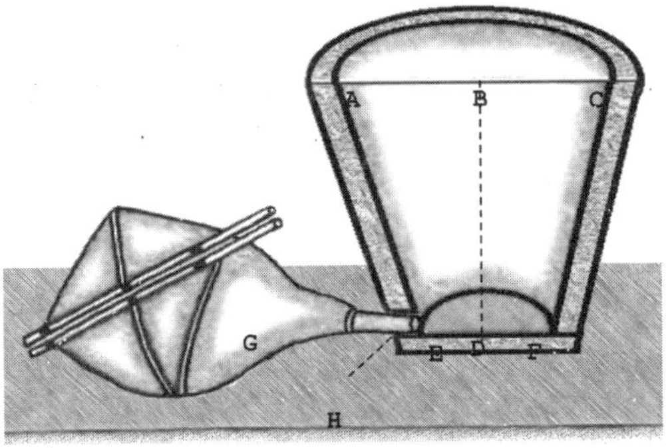

- *"dvādaśāṅgulakotse* the distance betweer points A, C and B, D *aṅgulas* i.e. 16 inches/ cms.
- *"sā budhne caturaṅ* the distance between I F = 4 *aṅgulas* approximately 3 inc 7.84 cms.
- G - *bhastrikā*
- H – passage to blow in a

fig. 10: Tiryakpātanakṣṭhī

Several Sanskrit treatises having such information are scattered all over the country which are the repository of scientific wisdom. Based on some of these texts the modern science developed. The rests are not yet came to light. The fragrance of this wisdom need to be spread far and wide including every household in our sacred land. With a view to digging out the hidden treatises as well as to restore this Indian wisdom, which is presently available in the form of manuscripts, the Government of India established the National Mission for Manuscripts in February 2003 and trying to document all such rare texts. It is observed that till date about 38,000 Āyurvedic manuscripts including Vṛkṣāyurveda and Mṛgāyurveda are already documented and out of which 1,200 (1,25,000 pages) *siddha* manuscripts are already digitised. It is expected by the Mission to complete this documentation work within next couple of years and thereafter the soft copies of all texts will be made available to the scholarly world.

References

Satpute, Ashok D., 2003, *Rasaratnasamuccaya*, Delhi: Chaukhamba Sanskrit Pratisthan.

Vaidyopādhyāya, Rāmaprasāda (ed.), 2000, *Rasendrapurāṇa*, Mumbai: Khemaraja Shrikrishnadas.

Vaiśya, Śyāmasundarācārya (Rasāyanaśāstrī), 1915, *Rasāyanasāra*, Kashi: Rasāyanaśālā.

Atri, Maharṣi, *Atri Saṁhita.*

Deshmukh, C. D. (tr.), 1981, *Amarakośa*, New Delhi: Utpal Publishing House.

Jośī, Veṇīmādhavaśāstri (Āyurvedācārya) and Nārāyaṇahari Jośī (Āyurveda Viśārada), 1968, *Āyurvedāya-śabdakośa (Sanskrit-Marathi-Sanskrit)*, Lakṣmaṇaśāstrī Jośī (Tarkatīrtha), Mumbai: Mahārāṣṭra Rājya Sāhitya Āṇi Saṃskṛti Maṇḍala.

Abbreviations

a. ko. — *Amarkośa*

a. saṁ — *Atri Saṁhitā*

ra.ra.sa — *Rasaratnasamuccaya*

ra. pu. — *Rasendrapurāṇa*

ra. sā — *Rasāyanasāra*

11

High Technology in Ancient Sanskrit Manuscripts

C.S.R. Prabhu

Bhāradvāja's *Vimānaśāstra* deals with advanced metallurgy, material science, machine design, mechanical engineering and rocketry. The text describes detailed procedures in the preparation of several hundreds of materials such as alloys and glasses which are unknown to modern science. Most of these materials can be reproduced in the laboratory even now. Some of the principles of metallurgy which are brought out in the textual description indicates an advanced development in technology and engineering. About 31 machines (*yantra*s) are also described with their construction procedures. Experimental investigation has been conducted for the materials part of the text. Several materials can be reproduced in the laboratory. By investigation it was found that they have special properties which are not available in any known materials of modern times. One machine *vakra prasāraṇa yantra* was reproduced as a working model and is found to be novel gear mechanism with 16-gear wheels. In addition, *agni stambhana*, a fire proofing spray, *anāhāra*, a food substitute, have also been produced. Patents are also being obtained for some of these items.

Introduction

THIS paper aims at presenting the preliminary results of the study and investigation on a few rare ancient Indian scientific Śāstras in Sanskrit. While there is a general opinion that at some point in the prehistoric or proto-historic times India had a highly developed technological society, there has been till now no real corroboration of this opinion.

The recent study and experimental investigations of a few rare and obscure Sanskrit works of scientific nature have provided some basis, especially in the fields of chemistry, metallurgy and material science that there could have indeed existed an advanced technological status of the society, at least in these fields, at some point of time in India's pre- or proto-history.

This paper attempts to present these initial results which include the procurement and decipherment of manuscripts and also the experimental laboratory preparation of the deciphered materials. A working model of a machine *vakra pṛasāraṇa yantra* has been developed.

History and Background

During January 1991 a set of Sanskrit manuscripts have been procured from a source at Bangalore. While most of the manuscripts procured were never published, limited publication was done for (a) *Vimānaśāstra* and (b) *Aṁśubodhinī*, both ascribed to Maharṣi Bhāradvāja.

Based on the study and investigation performed till now, the following history, background and origin of the system of Ancient Sanskrit Śāstras have been identified:

(i) An integrated system of ancient scientific Śāstras has been identified. This system is based on an integrated framework of concepts and principles unique and characteristic to ancient Indian literature including Vedic, Purāṇic and Āyurvedic components. These concepts have not much in common with modern science. However, the concepts can be tested only by independent physical verification. Since theoretical models of physical phenomena can be varied with time, the conceptual framework or model may not affect the physical phenomena themselves or the experimental results thereby achieved on empirical basis.

The Śāstras covered in this system run into over 100 in number with subjects as wide ranging as chemistry, metallurgy, engineering, architecture and medicine (human, veterinary and plant). We will be presently dealing with only the specific set of Śāstras which have been located by us during 1991, though there may be many other sources of Śāstras traditionally or otherwise which have not been dealt by us.

(ii) While only a few of the total system of Śāstras have actually been orally delivered and subsequently recorded into writing around CE 1912, more than twenty of them are supposed to have been actually available physically with the main source, namely the late Pandit T. Subbaraya Sastry of Anekal (near Bangalore) during the period of CE 1875 to 1930.

(iii) Pandit Subbaraya Sastry was apparently only a medium for oral delivery from his memory (which he reportedly acquired from his *guru*) of the Śāstras which were written down either in parts or in whole by others (as G. Venkatachala Sharma of Bangalore) who acted only as scribes.

About 20 Śāstras were orally delivered by him in parts and handwritten manuscripts (on old paper) were produced during 1911-40 as indicated by the dates recorded by the scribe (though he himself had access to them even during 1865-1911).

Some extracts of the Śāstras were also printed in a few articles in journal *Bhautika Kalānidhi* by late Shri B. Suryanarayana Rao (grandfather of Shri B.V. Raman of Bangalore who displayed them also).

(iv) While reportedly Pandit Subbaraya Sastry had access to these Śāstras through his *guru mahārāj* (who was also a *yogī*) the exact means and channels of acquisition have not yet been determined clearly. Possibilities of yogic meditation, trance or other extra-normal states being the cause for this cannot be ruled out, apart from regular memorisation for oral delivery.

(v) The small set of Śāstras actually delivered has quoted extensively from various other texts, including dictionaries, attributed to various authors of Vedic, post-Vedic periods on varied subjects of scientific nature. None of these texts have been located from any other sources, though the subject matter covered in them can be traced to be common to many Sanskrit works on related subjects, e.g. Āyurvedic texts and Nighaṇṭus (dictionaries).

(vi) Notwithstanding the unclear origin of these texts (for which reason many persons claimed that these Śāstras are not authentic or genuine) some of the contents of these texts have been investigated in terms of physical experimentation by trying out in the laboratory, the given formulae for the preparation of materials as alloys, glasses, ceramics, etc. as described in these texts. The decipherment process was primarily involved in tracing the synonyms of the words used for the input ingredients for making various materials by using Āyurvedic sources as dictionaries.

List of Procured Manuscripts of Ancient Scientific Śāstras in Sanskrit

	Tittle	Author ascribed
(i)	*Vimānaśāstra* (or *Vaimānika Prakaraṇam*)	Maharṣi Bhāradvāja
(ii)	*Aṁśu Bodhinī*	- do -
(iii)	*Kṛtakavajra Nirṇaya* (of *Ratna Pradīpikā*)	Not clear

(iv)	*Jalatattva Prakaraṇa*	- do -
(v)	*Apatattva Upanyāsa*	- do -
(vi)	*Rājya Tantra*	Maharṣi Yājñavalkya

The Decipherment Process

Among the various manuscripts procured, the following were mainly studied:

(i) *Vimānaśāstra* (or *Vaimānika Prakaraṇam*, a chapter of Bhārdvāja's *Yantra Sarvasva*)

(ii) *Aṁśubodhinī*

(iii) *Kṛtakavajra Nirṇaya*

The decipherment process centrally concerned itself on the identification of practicable formulae for chemical preparations (in the laboratory) from the corresponding Sanskrit descriptions and recipes for the preparation of the materials as alloys, glasses and ceramics, as per the ancient methods in terms of notions, concepts and framework which were unique to the Śāstras. The only common grounds with modern times were the input ingredient materials, which after decipherment could be identified either as equivalents of simple inorganic materials in the laboratory (as metals as copper, lead, etc.) or alternatively complex organic materials from the nature such as herbs, roots, gums, resins, barks and mineral ores largely known to the Āyurveda system, as is practised today, both by Āyurvedic physicians and Āyurvedic chemists and pharmacists.

The decipherment process which was quite tedious and complex, primarily involved finding the equivalent modern Sanskrit words for names of input ingredients for preparations of various materials. Since most of the words used in the manuscripts were quite archaic they have gone out of use in more recent (classical) Sanskrit, the Āyurvedic Nighaṇṭus or dictionaries helped to a partial extent in giving the more recent equivalents for such words. However there are quite a number of words for which no more recent equivalents have yet been found. The search efforts are still in progress.

The second part of the effort lies in finding the equivalent Indian language word for Sanskrit word, i.e. Hindi or Telugu equivalent which can be used for actual identification of herb or mineral ore in the market or in the nature in general. In this connection substantial help was obtained from Āyurvedic physicians and chemists.

The third step is procuring the identified materials from natural sources (e.g. mineral ore), a task sometimes becoming a very difficult, as some of the materials are vary rare and may not be available in the modern times.

The fourth step involves the actual preparation of the described material (as an alloy or a ceramic or glass) by mixing the identified input ingredients in the proportion given in the Sanskrit original.

At this stage the modern chemistry laboratory was utilised and the melting and cooling procedures were performed according to the directions given in Sanskrit manuscripts.

The mixing proportions were given in terms of relative units in weight. The units of temperature used were in *kakṣya*s and the exact interpretation of *kakṣya* is not yet known, though roughly it has been equated (as 1 *kakṣya* = 12.5 °C) at low temperatures (this scale may not be linear at higher temperatures).

The procedure for heat treatment was also given in Sanskrit sources in terms of either sudden pouring or gradual cooling or slow pouring to produce various effects and different properties.

As regards the equipment to be used, in all experiments only the modern laboratory equipment was used, though descriptions in Sanskrit were varied as various types of crucibles (*mūṣās*), bellows (*bhastrīs*) and furnaces (*kuṇḍas*) of ancient times (which are not available now).

Status Report

Ancient Indian texts and manuscripts pertaining to science and technology have been studied with an aim to decipher and decode formulae for making new materials as alloys, ceramics and glasses. A formula for making protein rich food extract from common Indian grasses also has been deciphered.

The following 20 formulae for new materials consisting of special alloys, ceramics and glasses have been deciphered and some of them were actually produced based on the formulae from the ancient Sanskrit texts of *Vimānaśāstra* (quoting Lohatantra), *Aṁśubodhinī*, *Kṛtakavajra Nirṇaya*, etc. based on the dictations of the texts and formulae by the Late Pandit Subbaraya Sastry of Anekal (CE 1855-1940).

1. ***Tamogarbha loha*** **:** A lead alloy capable of light absorption

Status **:** Already produced in the laboratory, light in weight, black in colour, found to be resistant to acids.

Displayed high level of absorption for laser light (from red Ruby laser — as observed by Prof. Robert Anderson of San Jose State University during his visit to India in December 1991). Some chemical and other properties found to be unique — patentable new alloy. A laboratory test done in 1996 in Physics Department of Osmania University, Hyderabad, indicated laser absorption characteristics up to 79 per cent of incident light from a laser. (this alloy was used in *tamo yantra* in the *Vimānaśāstra* for the purposes of absorption of light escaping from a photochemical reaction which resulted in absorption of light, thereby generating "darkness").

2. ***Pañca loha (not the well-known Pañcaloha for making idols)*** : A copper alloy which is highly malleable and also highly corrosion resistant to salt (NaCL)

Status : Already produced and characterised to possess:

(a) Golden yellow colour (described in the Sanskrit text as *hema varṇam* or golden colour).

(b) Corrosion resistance to moisture and salt water (displayed 2 weight loss of only about 0.00335 mg/dm/day in 3 per cent NaCL solution).

(c) High machinability and on microstructure analysis found to be single phase alloy with high malleability (described in Sanskrit as *mṛdulam* or "soft").

(d) Characteristics, composition and properties found to be not listed in ASM Reference (1988) and therefore patentable new alloy.

3. ***Arāra Tāmra*** : A copper alloy zinc, lead and iron of light absorption.

Status : Already produced and characterised to possess:

(a) Golden yellow to reddish tinge (described in Sanskrit text as *hema varṇam* or golden colour).

(b) Brittle, light and hard on microstructure analysis found to be two phase alloy.

(c) Very hard (Young's modulus 16.9) (described in Sanskrit text as *Dṛḍham* or "strong").

(d) Characteristics, composition and properties found to be not listed in ASM Reference (1988) and therefore patentable new alloy.

4. ***Capala grāhaka* (ceramic)** : A fine porcelain type of ceramic.

Status : Already produced and characterised to be resistant to all acids and alkalis.

5. ***Capala grāhaka* (glass)** : A soft glass (of low temperature melt).

***Status* :** Already produced and characterised to be resistant to acids and alkalis.

Refractive index found to be 1.614. (highest known among soft glasses made at low temperatures).

6. ***Ravi śakti apakarṣaṇa darpaṇa (glass)* :** A special glass concentrating (visible) light energy in sunlight.

***Status* :** Already produced and study of optical properties is not yet done.

7. ***Uṣṇa śakti apakarṣaṇa darpaṇa* (glass) :** A special glass for concentrating the heat energy in sunlight.

***Status* :** Fully deciphered and to be produced in the laboratory.

8. ***Badhira loha* :** A sound proof alloy.

***Status* :** Fully deciphered and to be produced in the laboratory.

9. ***Vidyut darpaṇa* :** A special glass that has capability to neutralise electrical discharges as lightning.

***Status* :** Fully deciphered and to be produced.

10. ***Raja loha* :** A high-heat-absorbing alloy used for the bodies of various flying crafts.

***Status* :** Fully deciphered and to be produced in the laboratory.

11. ***Rudanti maṇi* :** A special material.

***Status* :** Fully deciphered and to be produced in the laboratory.

12. ***Rutika maṇi* :** A special material.

***Status* :** Fully deciphered and to be produced in the laboratory.

13. ***Abhra mṛd darpaṇa* :** A special mica glass.

***Status* :** Fully deciphered and to be produced in the laboratory.

14. ***Sunda mṛt kāca* :** A special glass.

***Status* :** Fully deciphered and to be produced in the laboratory.

15. ***Piṅgala adarśa*** : A special glass.

Status : Fully deciphered and to be produced in the laboratory.

16. ***Somanka loha*** : A special alloy.

Status : Fully deciphered and to be produced in the laboratory.

17. ***Ravi śakti apakarṣana darpaṇa*** : A special glass with solar heat collecting properties.

Status : Fully deciphered and to be produced in the laboratory.

18. ***Hatakasya loha*** : A copper alloy with golden appearance.

Status : Fully deciphered and to be produced in the laboratory.

19. ***Vāta stambhana loha*** : A copper, iron, lead alloy.

Status : Fully deciphered and to be produced in the laboratory.

20. ***Ghaṇṭarava loha*** : An alloy that has high sensitivity to different types of sounds.

Vakra Prasaraṇa Yantra : Design and Fabrication

The *arāra tāmra* (item no. 3 above) was described to be basic alloy material for making the cylindrical frame enclosure for the *vakra prasaraṇa yantra* given in *Vimānaśāstra*. This machine has been reproduced as a working model at M/s MTE Industries, Hyderabad. This *yantra* is an advanced machine. This machine was stated to enable sharp turns, circular motion and reverse turn of *vimāna* or any vehicle. This machine is found to be a gear mechanism with 16-gear wheels in a conical design. This has one input and two outputs. One of the two outputs moves in the same direction as the input while the other output moves in the opposite direction as the input. The speed of the second output also is higher than the first output. Multiple outputs can be taken out. Such a mechanism is unknown and new in today's machine design. Patent application is being made for this *yantra*.

Low Cost Protein Rich Food from Grass

In addition to the above materials, a formula for producing a protein rich food extract (powder) from specific Indian grasses is also deciphered. The Central Food Tech. Research Institute, Hyderabad, has certified that the powder extracted from the specified grasses shows about 13 per cent protein content. Other tests from nutrition and medical angles are to be taken up. This activity is aimed at producing low-cost protein rich food products (as powder, biscuits, malt, etc.) based on this formula.

Agni Stambhana (Fire Resistance)

In addition to materials produced or deciphered as above, a technique for preventing and resisting fire and burning is also developed. Two techniques/solutions have developed for *agni stambhana*:

(a) for preventing burning of inflammable objects as paper, cloth and wood

(b) for preventing burning of human body.

It has been noted that in both the above cases the fire will not be allowed to be caught (or burning to start) even after continuous exposure to flame for up to 30 seconds. (Normally fire catches any inflammable material with 0.5 seconds and any moist material within 3 dampened with this liquid do not catch fire and can also be used to put off fire or escape unburnt in fire even after long exposure to flame.) A patent has been obtained for this invention from Madras Patent Office.

8. Anāhāra (Avoiding Food)

A recipe for a special type of biscuit has been developed. This biscuit when consumed up to 50 g dose can help overcome hunger and skip a meal for about 3-4 hours. Upon medical trial, it was found very useful in obesity and diabetic cases.

(Note: Items 7 & 8 are developed independent of Vimānasāstra.)

Details of Individuals and Organisations Involved and Acknowledgement

Acknowledgement with deep gratitude are due to the following individuals and organisations who have been contributing in various capacities and roles for the success of this project:

- Dr B.G. Siddharth, Director, Birla Science Centre, Birla Science Centre, Hyderabad [*Role*: Committee member (of the three-member committee for investigation into ancient Indian science and technology) and had played crucial role in arranging various resources for the project.
- Late Dr M.C. Ganorkar Director, BISR, Hyderabad and Ranga Madhavan, (then Doctoral student), Birla Institute of Scientific Research (BISR), Hyderabad [*Role*: Committee member (of the three-member committee for and investigation into ancient Indian science and technology) and had played a crucial role in terms of conducting various experiments and all activities related to chemistry in terms of preparation of materials as per formulae in the Śāstras, etc. and also testing for chemical properties of materials produced in the laboratory BISR, Hyderabad.
- B.K. Chandrasekhar, Sr. Manager, CTI Bangalore, BHEL-CTI, [*Role*: Assistance in melting at high temperature].
- R.B. Subramanian, Project Director, DMRL [*Role*: For assisting with providing facilities in DMRL].
- Dr Ram Niwas Sharma, Dr Surendra Sharma, Dr K.G. Sharma, Punarvasu Arogya Kendra, Secunderabad [*Role*: For assisting with Āyurvedic information in decipherment of various formulae].
- Dr B.V.S. Subba Rao [*Role*: For assisting with ceramics-related information and also interacted with Indian ceramics society].
- Dr Y.V. Ramana, NGRI [*Role*: For performing mechanical tests on alloys produced at NGRI, Hyderabad, and producing valuable data].
- Sri S.R. Sarma [*Role*: For advice from time (Retired Director to time in the process of Mines & Geology, of decipherment. Govt. of A.P.)].
- Head Palace Manuscript Library, Jaipur, Maharaja [*Role*: For permitting to use the library of Ancient manuscripts in Sanskrit].
- Head, IRDE [*Role*: For offering Dehradun test facilities].

- Sri. Ch. Sathi Reddy, Head, Industries, Dindigul, Hyderabad [*Role*: M/s MTE, For arranging for the design, casting and fabrication of *vakra prasaraṇa yantra*].
- Shri Khuleesh, Kothari, Jaipur [*Role*: For offering support to this work].
- Prajna Bharati, Hyderabad [*Role*: For sponsoring and arranging resources].

Conclusion

This investigation provides the basis for identifying the veracity of the description given in source of the Sanskrit texts dealing with materials and also machine design. While it is not clear how these texts originated or whether they are authentic, the experimental approach only attempted in finding out the validity of the description of preparation of some special materials such as alloys, glasses, ceramics, etc. and also one machine *vakra prasāraṇa yantra*. Even though some persons have hinted at the whole text being a modern work, composed by the oral deliverer himself, the experimental results uphold the veracity of the textual contents and also indicate lack of availability of these materials and machines in their exactness in modern times, thereby hinting at an ancient historical origin.

12

Environmental Science
Vedic Perspective

Shashi Tiwari

Vedic vision on environment is well defined in one verse of the *Atharvaveda* where three coverings of our surroundings are referred as *chandāṁsi*. Vedic Āryans studied nature's drama very minutely. They realised instinctively that action, movement, creation, change and destruction in nature are the results of forces beyond men's control. Natural aspects or energies are glorified by the Vedic seers as deities. Yāska has classified these gods among three groups, i.e. upper, middle and lower according to the division of the universe. *Ṛta* is the name given to Cosmic Law which governs everything in the universe. Varuṇa is depicted as the Lord of this universal order. The Bhūmi-Sūkta of *Atharvaveda* denotes environmental consciousness of Vedic people especially in reference to earth. The hymn talks about the different energies which are generated from the body of the earth. According to *Ṛgveda* water as a part of human environment occurs in five forms. Vedic verses also describe characteristics and significance of air for living beings. Animals and birds are regarded as a natural part of human environment. Oṣadhi-Sūkta of *Ṛgveda* addresses to plants and vegetables as mother. The concept of Vedic *yajña* seems to be a major principle of environmental science in its broader sense. The Vedic view to live in harmony with environment was not merely physical but was far wider and much comprehensive.

Introduction

VEDAS are the first texts in the library of mankind. They are universally acknowledged to be the most precious Indian heritage. The antiquity to the Vedic civilisation is debated to a great extent but indeed there is no civilisation known to humanity with such antiquity as Vedic Āryan civilisation. The so-called Āryans would have originated in the Āryāvarta. N. J. Lockyer has declared:

> The Vedas, in fact, are the oldest books in which we can study the first beginnings of our language and of everything which is embodied in all the languages under the sun.[1]

The Vedas deal with knowledge, the knowledge of all sorts. They cover knowledge both physical and spiritual. They are source of all knowledge according to *Manusmṛti*.[2] Especially the Vedic views revolve around the concept of nature and life. The visions of the beauty of life and nature in the Vedas are extremely rich in poetic value. Perhaps nowhere else in the world has the glory of dawn and sunrise and the silence and sweetness of nature, received such rich and at the same time such pure expression. The symbolical pictures projected there remain close to life and nature. The most authoritative among the four Vedas is called the *Ṛgveda*. Each Vedic verse has one or more sages (*ṛṣi*s) and deities (*devatā*s) associated with it. Generally, *ṛṣi*s are supposed to be the recipient of knowledge revealed in the verses and *devatā*s are supposed to be the gods in whose praise verses are revealed.

The oldest and simplest form of nature-worship finds expression in Vedic texts. Many scholars have come to the conclusion that the Vedas are primarily concerned with cosmology, however, they are not in a position to show that Vedic cosmology has the solutions to the most difficult problems of modern cosmology.[3] Some say, like dramas are

1. N.J. Lockyer, *The Dawn of Astronomy*, Massachusetts: Institute of Technology, USA, p. 432.
2. *sarvaṁ vedāt prasidhyati — Manusmṛti*, 2.7.
3. Raja Rammohun Roy, 1999, *Vedic Physics, Scientific Origin of Hinduism*, Toronto: Golden Egg Publishing, p. 6.

played to remember history, the process of various *śrauta yajña*s describes the science of cosmology.[4]

The Vedic hymns are full of statements, ideas and unusual images which contain truths of all sciences. Here, knowledge is couched in symbolic language and unless the symbols are decoded, the real purport of the *mantra*s cannot be understood. The only point is that Vedas need to be studied and interpreted, not in a pedantic manner, but in their proper perspective and in relevant context. The tripartite model of knowledge at the basis of the hymns helps in their understanding. Generally indication of most of the principles is there in their earliest form. Often expressions of ideas are enveloped with the shade of symbolism. The approach of Vedic seers is truly comprehensive. They do not visualise in parts. They do not elaborate subjects as is done in current education. But at the same time, grandeur and brevity of the Vedas are not found in the disciplines of modern science. The Vedas and disciplines of modern science are rather complementary and not contradictory. If modern science is seen or read through Vedic eyes, the students will be much benefited. Students of science may search the earliest of the ideas about any discipline in the Vedic literature.

In recent days, environmental science and ecology are disciplines of modern science under which study of environment and its constituents is done with minute details. As science, they are established in twentieth century, but their origin can be seen long back in the Vedic and ancient sanskrit literature. The concepts of environment differ from age to age, since it depends upon the condition, prevalent at that particular time. In this paper, an effort is made to find out the awareness of ancient Indian people about the environment.

4. Yudhishthira Mimamsaka, 1976, *Vaidika Siddhānta Mimāṁsā*, Sonepat, p. 40.

As Sanskrit literature is so wide we refer here mainly to Vedic texts, particularly the Vedic Saṁhitās.

The Environment (Protection) Act, 1986 defines the environment as follows:

> Environment includes water, air and land and the interrelationship which exists among and between water, air and land and human beings, other living creatures, plants, micro organisms and property.[5]

From the above definition, it can be briefly said that environment consists of two components, namely biotic (living organisms) and abiotic (non-living materials) factors. The living organisms can be grouped into three types — those living mainly on land, in water and in air. The non-living materials of the environment are land, air, water, property, etc.

In modern Sanskrit, the word *paryāvaraṇa* is used for environment, meaning which encircles us, which is all around in our surroundings. But in the *Atharvaveda* words equivalent to this sense are used; such as *vr̥tāvr̥tā*,[6] *abhivārāḥ*,[7] *avr̥taḥ*,[8] *parivr̥tā*,[9] etc. Vedic view on environment is well defined in one verse of the *Atharvaveda* where three coverings of our surroundings are referred as *Chandāṁsi*:

> Wise utilise three elements variously which are varied, visible and full of qualities. These are water, air and plants or herbs. They exist in the world from the very beginning. They are called as *Chandāṁsi* meaning "coverings available everywhere".[10]

---

5. A.R. Pancamukhi, 1998, *Socio-Economic Ideas in Ancient Indian Literature*, Delhi: Rashtriya Sanskrit Sansthan, p. 467.
6. *vr̥tāvr̥tā* — *Atharvaveda* 12.1.52.
7. *abhīvāraḥ* — ibid., 1.32.4.
8. *āvr̥tāḥ* — ibid., 10.1.30.
9. *parīvr̥tāḥ* — ibid., 10.8.31.
10. *trīṇi chandāṁsi kavayo vi yetire pururūpaṁ darśataṁ viśvacakṣaṇam. āpo vātā oṣadhayastāny ekasmin bhuvana ārpitāni* | | — ibid., 18.1.17

It proves the knowledge of Vedic seers about the basic elements of environment.

According to one indigenous theory established in the Upaniṣads, the universe consists of five basic elements — viz. 1. earth or land, 2. water, 3. light or lustre, 4. air, and 5. ether.[11] The nature has maintained a status of balance between and among these constituents or elements and living creatures. A disturbance in percentage of any constituent of the environment beyond certain limits disturbs the natural balance and any change in the *natural balance* causes lots of problems to the living creatures in the universe. Different constituents of the environment exist with set relationships with one another. The relation of human being with environment is very natural as he cannot live without it. From the very beginning of creation he wants to know about it for self-protection and benefit.

Vedic Approach to Environment

The Vedic Āryans were children of nature. They studied nature's drama very minutely. Sandstorm and cyclone, intense lightning, terrific thunderclaps, the heavy rush of rain in monsoon, the swift flood in the stream that comes down from the hills, the scorching heat of the sun, the cracking red flames of the fire, all witness to power beyond man's power. The Vedic sages felt the greatness of these forces. They adored these activities. They appreciated these forces. They worshipped and prayed them due to regard, surprise and fear. They realised instinctively that action, movement, creation, change and destruction in nature are the results of forces beyond men's control. And thus they attributed divinity to nature.

---

11. *imāni pañcamahābhūtāni pṛthivī, vāyuḥ ākāśaḥ āpa jyotīṁṣi* — *Aitareya Upaniṣad* 3.

DIVINITY TO NATURE

Ṛgvedic hymns could be divided into many parts, but their main part belongs to natural hymns, the hymns related with natural forces. Yet Vedic gods are explained in different ways by the scholars of India and West, but speaking generally, the hymns addressed to deities (*devatā*) are under the influence of the most impressive phenomenon of nature and its aspects. The word *devatā* means divine, dignity which is bright, strong, donor, and powerful. In these hymns we find prayers for certain natural elements such as air, water, earth, sun, rain, dawn, etc. The glorious brightness of the sun, the blaze of the sacrificial fire, the sweep of the rainstorm across the skies, the recurrence of the dawn, the steady currents of the winds, the violence of the tropical storm and other such natural energies, fundamental activities or aspects are glorified and personified as divinities (*devatā*). The interaction with nature resulted in appreciation and prayer but, indeed, after a good deal of observation. Attributes assigned to deities fit in their natural forms and activities, as Soma is green, fire is bright, air is fast-moving and sun is dispenser of darkness. The characteristics of these forces described in the verses prove that Vedic seers were masters of natural science.

In Vedic view this world consists of Agni, i.e. fire or heat, and Soma, i.e. water.[12] Sun (Sūrya) is the soul of all which is moving and also of which is not moving.[13] Indra is most powerful god who kills Vṛtra, the symbol of cloud to free waters. Vṛtra means one who covers and is derived from the root *vṛ*, to cover. R.R.M. Roy opines that the main force of expansion in the Vedic cosmology is Indra, and his chief adversary, the main force of contraction, is Vṛtra.[14] Maruts are Indra's associates. Vedic seers pray boldly to these natural

12. *agnisomātmakaṁ jagat.*
13. *sūrya ātmā jagatastasthuṣaśca* — *Ṛgveda* 1.115.1.
14. Raja Rammohun Roy, op. cit., p. 58.

forces and aspects for bestowing plenty and prosperity on them. Aditi is praised as *devamātā,* the mother of all natural energies and she symbolises the Nature.

A famous geologist S.R.N. Murthy has written on the earth sciences in the Vedas. He has somehow a different opinion about Vedic gods and hence states, "the natural geological aspects have been described as Indra, Agni, Vāyu, Varuṇa, Uṣas, etc."[15]

COSMIC ORDER "ṚTA" AND VARUṆA

In the Vedas, the order of the Universe is called *ṛta. Ṛta* reduces chaos to cosmos, and gives order and integration to matter. It also gives symmetry and harmony in the environment. Hence the conception of *ṛta* has an aesthetic content too; it implies splendour and beauty. It is for this reason that the Vedic gods, upholding *ṛta,* are all lawful, and beautiful and good. Their beauty is a significance attribute.

Ṛta is defined variously by scholars in different Vedic contexts, but in general sense it has been elaborated as great "cosmic order" which is the cause of all motion and existence and keeps world in order. No one can ignore it,[16] even gods are abided by the *ṛta,* they are born of *ṛta.* It is controlling and sustaining power. It sustains sun in the sky.[17] *Ṛta* as Universal Law governs everything in the cosmos. The whole of the manifested universe is working under *ṛta.* S.R.N. Murthy assumes it as a law of gravitation in simple form. According to H.W. Wallis

> The principle of the order of the world, of the regularity of cosmic phenomena, was conceived by the *ṛṣi*s to have existed

15. S.R.N. Murthy, 1997, *Vedic View of the Earth,* New Delhi: D.K. Printworld, p. 12.
16. *ṛtaṁ nātyeti kiñcana* — *Taittirīya Brāhmaṇa* 1.5.5.1.
17. *satyenottabhitā bhūmiḥ sūryeṇottabhitā dyauḥ.*
ṛtenādityāstiṣṭhanti divi somo adhi śritaḥ — *Ṛgveda* 10.85.1.

> as a principle before the manifestation of any phenomena. The phenomena of the world are shifting and changeable, but the principle regulating the periodical recurrence of phenomena is constant; fresh phenomena are continually reproduced, but the principle of order remains the same; the principle, therefore, existed already when the earliest phenomena appeared.[18]

In the Vedas, Varuṇa is depicted as the Lord of *ṛta*, the universal natural order. He is sovereign god, great king, law-maker and ruler of cosmos and even of the gods. Basically, he is regarded as the Lord of water and ocean but chiefly he controls and keeps the world in order. From his throne on high he looks down upon all that happens in the world, and into the heart of man.[19]

> By the law of Varuṇa heaven and earth are held apart. He made the golden swing, the sun to shine in heaven. He has made wide path for the sun. By his ordinances the moon shining brightly moves at night, and the stars placed up on high are seen at night but disappear by day. He causes the rivers to flow. As a moral governor Varuṇa stands far above any other deity.[20]

Thus, the concept of Varuṇa represents the consciousness of Vedic seers in respect of controlling and balancing the natural forces in environment.

DIVISION OF UNIVERSE

Vedic seers have a great vision about universe. The universe is made on scientific principles, and that's why it is well measured. The universe consists of three intertwined webs,

---

18. H.W. Wallis, 1999, *The Cosmology of the Ṛgveda*, Delhi: Cosmo Publications, pp. 94-95.
19. Ibid., pp. 91-101; *Ṛgveda* 1.25.7, 9, 11.
20. A.A. Macdonell, 1965, *A History of Sanskrit Literature*, Delhi: MLBD, pp. 61-62.

pṛthivī, antarikṣa and *dyau.* Vedic scientists divided even the length in three calling them upper, medium and lower. The tripartite division of the universe into three regions — *pṛthivī* (the earth), *antarikṣa* (the aerial or intermediate region) which is between heaven and earth, and *dyau* (the heaven or sky) is very well established in the Vedic literature. *Pṛthivī* can be given a scientific name "observer space". It is our space, the space in which we live and die, whatever we can see and observe. From one end of the universe to the other end is the expanse of *pṛthivī*, and that is what the name *pṛthivī* means: the broad and extended one. *Dyau* can be termed "light space" because light propagates in this space. *Antarikṣa* can be termed as "intermediate space" as this space exists in between observer space and light space. A verse from the *Yajurveda* states that the division of universe was done on a subtle level, and not on gross level.[21] The Vedic sages had the capability of looking at such a subtle level, which is beyond the reaches of modern science. Here, in reference to environmental study, we regard the division of the universe as the most important concept of the Vedas.

Though a large number of gods are described in the hymns, and it is very difficult to arrange them in different classes, but Yāska in his *Nirukta* talks about three gods: Agni in earth, Vāyu or Indra in atmosphere and Sun in heaven. Each one of them is known by various names depending on the different actions performed.[22] These three gods are three major forms of energy, fire on earth, air in intermediate space and light in upper region. Other energies of those regions are related to or under them. So generally gods are classified in three groups called upper, middle and lower, and, therefore, provide a system to study atmosphere and its all aspects. Regarding global harmony, Vedic seers always pray for the welfare of all creatures and all regions.

21. *Yajurveda*, 7.5.

22. *Nirukta*, 7.5.

Concept of the Earth (Pṛthivī)

The concept of the form of the earth in the *Ṛgveda* is most fascinating. It is mostly addressed along with the heaven into a dual conception (*rodasī, dyāvāpṛthivī*). There is one small hymn addressed to *pṛthivī,* while there are six hymns addressed to *dyāvāpṛthivī. Pṛthivī* is considered the mother and *dyau* is considered the father in the Vedas, and they form a pair together. One of the most beautiful verse of the *Ṛgveda* says: "Heaven is my father, brother atmosphere is my navel, and the great earth is my mother."[23] Heaven and earth are parents : *mātarā, pitarā, janitārā*[24] in union while separately called as father and mother. They sustain all creatures. They are parents of all gods. They are great (*mahā*) and widespread. Earth is described as a goddess in *Ṛgveda*. In the *Atharvaveda* the earth is described in one hymn of 63 verses. This famous hymn called as Bhūmi-Sūkta or Pṛthivī-Sūkta indicates the environmental consciousness of Vedic seers. The seers appear to have advanced understanding of the earth through this hymn. She is called *vasudhā* for containing all wealth, *hiraṇyavakṣā* for having gold bosom and *jagato niveśanī* for being abode of whole world. She is not for the different races of men alone but for other creatures also.[25] She is called *viśvambharā* because she is representative of the universe. She is the only planet directly available for the study of the universe and to realise the underlying truth. This is wide earth which supports varieties of herbs, oceans, rivers, mountains, hills, etc. She has at places different colours as dark, tawny, white. She is raised at some place and lowered at some places. The earth is fully responsible for our food and prosperity. She is praised for her strength. She is served day and night by rivers and protected by sky.

23. *dyaurme pitā janitā nābhiratra bandhurme mātā pṛthivī mahāyam* — *Ṛgveda* 1.164.33.
24. *mātarā, pitarā, janitārā* — *Ṛgveda* 1.159.160.
25. *tvaṁ bibharṣi dvipadaḥ tvaṁ catuṣpadaḥ* — *Atharvaveda* 12.1.15; 12.1.45.

The immortal heart of earth is in the highest firmament (*vyoma*). Her heart is sun.

> She is one enveloped by the sky or space and causing the force of gravitation. She is described as holding Agni. It means she is described as the geothermal field. She is also described as holding Indra, i.e. the geomagnetic field. The earth is described then as being present in the middle of the oceans (sedimentary rocks) and as one having magical movements.[26]

The hymn talks about different energies which are generated from the form of the earth. "O *pr̥thivī* ! thy centre, thy navel, all forces that have issued from thy body, set us amid those forces; breathe upon us."[27] Thus, the earth holds almost all the secrets of nature, which will help us in understanding the universe. She is invested with divinity and respected as mother: The earth is my mother and I am Her son."[28] The geographical demarcations on this earth have been made by men and not by nature.

Concept of Water (Āpaḥ)

Water is essential to all forms of life. According to *R̥gveda* the water as a part of human environment occurs in five forms:

1. Rain water (*divyāḥ*)
2. Natural spring (*śravanti*)
3. Wells and canals (*khanitrimāḥ*)
4. Lakes (*svayamjāḥ*)

---

26. S.R.N. Murthy, op. cit., p. 87.
27. *yatte madhyaṁ pr̥thivī yacca nabhyaṁ yāsta ūrjastanvaḥ sambabhūvuḥ. tāsu no dhehyabhi naḥ pavasva mātā bhūmiḥ putro 'ahaṁ pr̥thivyāḥ parjanyaḥ pitā sa u naḥ pipartu* — *Atharvaveda* 12.1.12.; R.T.H. Griffith, 1995, *The Hymns of the Atharvaveda*, Delhi: D.K. Publishers, vol. II , p. 95.
28. *mātā bhūmiḥ putro 'haṁ pr̥thivyāḥ* — *Atharvaveda*, 12.1.12.

5. Rivers (*samudrārthā?*)[29]

There are some other classifications also in the *Taittirīya Āraṇyaka*,[30] *Yajurveda*[31] and *Atharvaveda*[32] as drinking water, medicinal water, stable water, etc. *Chāndogya Upaniṣad* describes about qualities of water:

> The water is the source of joy and for living a healthy life. It is the immediate cause of all organic beings such as vegetations, insects, worms, birds, animals, men, etc. Even the mountains, the earth, the atmosphere and heavenly bodies are water concretised.[33]

The cycle of water is described. From ocean waters reach to sky and from sky come back to earth.[34] Rain-waters are glorified. The rain-cloud is depicted as Parjanya god.

The fight between Indra and Vṛtra is celebrated story from the *Ṛgveda*. It is explained in many ways. According to one view it is a fight for waters. Indra is called Apsu-jit or conquering the waters, while Vṛtra is encompassing them. Vṛtra holds the rain and covers waters and thus being faulty is killed by Indra through his weapon called *vajra*, i.e. (thunderbolt). The Indra–Vṛtra fight represents natural phenomenon going on in the aerial space. By the efforts of Indra all the seven rivers flow. The flow of water should not be stopped and that is desired by humanity.

The significance of water for life was well known to Vedic

---

29. *yā āpo divyā uta vā sravanti khanitrimā uta vā yāḥ svayaṁjāḥ — samudrārthā yāḥ śucayaḥ pāvakāstā āpo devīriha māmavantu* — *Ṛgveda* 7.49.2.
30. *Taittirīya Āraṇyaka* 1.24.1-2.
31. *Yajurveda* 22.25.
32. *Atharvaveda* 1.6.4.
33. *Chāndogya Upaniṣad* 7.10.1
34. *apaḥ samudrāddivamudvahanti divaspṛthivīmadhi ye sṛjanti* — *Atharvaveda*, 4.27.4.

seers. They mention: waters are nectars;[35] waters are source of all plants and giver of good health;[36] waters destroy diseases of all sorts;[37] waters are for purification.[38] It seems that later developed cultural tradition of pilgrimage on the riverbanks is based on the theory of purification from water. The ancient Indians knowing water as a vital element for life, were very particular to maintain it pure and free from any kind of pollution. The *Manusmṛti* stresses on many instances to keep water clean.[39] The *Padma Purāṇa* condemns water pollution forcefully saying, "the person who pollutes waters of ponds, wells or lakes goes to hell".[40]

Concept of Air (Vāyu)

The observer space is the abode of matter particles, light space is the abode of energy and the intermediate space *antarikṣa* is the abode of field. The principal deity of *antarikṣa* is Vāyu. *Jaiminīya Brāhmaṇa* quotes, "Vāyu brightens in *antarikṣa*". Field is another form of energy and, therefore, *Yajurveda* says, "Vāyu has penetrating brightness". The meaning of Vāyu is made clear in *Śatapatha Brāhmaṇa* in the following *mantra*, "Sun and rest of universe is woven in string. What is that string, that is Vāyu." This verse clearly shows that here Vāyu cannot mean air alone.[41]

Apparent meaning of Vāyu is air. The Vedic seers knew the importance of air for life. They understood all about air in the

---

35. *amṛtaṁ vā āpaḥ* — *Śatapatha Brā.* 1.9.3.7.
36. *āpaḥ viśvabheṣajīḥ* — *Ṛgveda* 1.23.20
37. *bheṣajīrāpo 'mīvacātanīḥ* — *Ṛgveda* 10.87.6
38. *pavitraṁ vā āpaḥ* — *Śatapatha Brā.* 1.1.1.1, *āpaḥ pavitramucyante* — *Nirukta* 5.6.2
39. *Manusmṛti* 4.56.
40. *sukūpānāṁ taḍāgānāṁ prapānāṁ ca paraṁtapa sarasāṁ caiva bhettāro narā nirayagāminaḥ* — *Padma Purāṇa*, 96.7.8.
41. Raja Rammohun Roy, op. cit., p. 84; *Jaiminīya Brā.* 1.192; *Yajurveda* 1.24; *Śatapatha Brā.* 8.7.3.10.

atmosphere and also about the air inside the body. The *Taittirīya Upaniṣad* throws light on five types of wind inside the body: *prāṇa, vyāna, apāna, udāna* and *samāna.* Air resides in the body as life.[42] Concept and significance of air is highlighted in Vedic verses. *Ṛgveda* mentions: "O Air! You are our father, the protector."[43] Air has medicinal values.[44] "Let wind blow in the form of medicine and bring me welfare and happiness."[45] Medicated air is the international physician that annihilates pollution and imparts health and hilarity, life and liveliness to people of the world. Hilly areas are full of medicated air consisted of herbal elements. Another verse describes characteristics of air:

> The air is the soul of all deities. It exists in all as life-breath. It can move everywhere. We cannot see it. Only one can hear its sound. We pray to air God.[46]
>
> Ancient Indians, therefore, emphasised that the unpolluted, pure air is source of good health, happiness and long life. Vāyu god is prayed to blow with its medicinal qualities.

Concept of Ether (Ākāśa)

Modern environmentalists discuss sound or noise pollution. There is a relation between ether and sound. The sound waves move in sky at various frequencies. Scientist could see the sky which exists only in the vicinity of earth, but *Taittirīya Upaniṣad* throws light on two types of ether, i.e. one inside the body and the other outside the body.[47] The ether inside the body is

---

42. *vāyurha vā prāṇo bhūtvā śarīramāviśat* — *Taittirīya Upaniṣad*, 2.4.
43. *uta vāta pitāsi naḥ* — *Ṛgveda* 10.186.2.
44. *ā vāta vāhi bheṣajam* — ibid., 1.37.2.
45. *vāta ā vātu bheṣajaṁ śaṁbhu mayobhu no hṛde* — ibid., 10.186.1.
46. *ātmā devānāṁ bhuvanasya garbho yathāvaśaṁ carati deva eṣaḥ*
ghoṣā idasya śṛṇvire na rūpaṁ tasmai vātāya haviṣā vidhema
— Ibid. 10.168.4.
47. *say a eṣo antarhṛdaya ākāśa?. suvarityasau* — *Taittiriya Upaniṣad* 1.6.1; 1.5.1.

regarded as the seat of mind. An interesting advice to the mankind is found in the *Yajurveda*:

> Do not destroy anything of the sky and do not pollute the sky. Do not destroy anything of *antarikṣa*.[48]

Sun shines in *dyuloka* and we get light from sky. The sunrays strengthen our inner power and are essential for our life. Thus importance and care for ether is openly mentioned in the Vedic verses.

Concept of Mind (Manas)

Many prayers are found in Vedas requesting the God to keep the mind free from bad thoughts, and bad thinking. In this regard the Śivasaṁkalpa-Sūkta of *Yajurveda* is worth mentioning.[49] Considering the havoc that the polluted minds may create, our ancient sages prayed for a noble mind free from bad ideas. The logicians recognise *manas* as one of the nine basic substances in the universe.[50] The mind is most powerful and unsteady. Although the study of mind does not appear directly under the contents of modern environmental science but in reference to cultural environmental consciousness of Vedic seers, we find many ideas discussed in Vedic literature on the pollution of mind and its precautions.[51]

Animals and Birds

Animals and birds are part of nature and environment. It is natural, therefore, that Vedic seers have mentioned about their characteristics and activities and have desired their welfare. *Ṛgveda* classifies them in three groups — sky animals like birds,

48. *dyāṁ mā lekhīrantarikṣaṁ mā hiṁsīḥ* — *Yajurveda* 5.43.
49. Ibid., 34.1-6.
50. *Tarkasaṁgraha*, 2.
51. Nandita Singhavi, 2004, *Vedoṁ Meṁ Paryāvaraṇa*, Jaipur: Sonali Publications, pp. 313-56.

forest animals and animals in human habitation.[52] All the three types of living creatures found in the universe have distance environment and every living creature has an environment of its own. But when we look from man's perspective all of them constitute his environment. There is a general feeling in the Vedic texts that animals should be safe, protected and healthy.[53] Domestic animals as well as wild animals along with human beings should live in peace under the control of certain deities like Rudra, Pūṣan, etc. Vedic people have shown anxious solicitude for welfare of their cattle, cows, horses, etc. The cow as the symbol of wealth and prosperity, occupied a very prominent place in the life of the people in Vedic times.[54]

Plants and Herbs (Oṣadhi)

The knowledge about the origin and significance of plants can be traced out from Vedic literature in detail. In *Ṛgveda* one Araṇyāni-Sūkta is addressed to the deity of forest.[55] Araṇyāni, queen of the forest, received high praise from the sage, not only for her gifts to men but also for her charm. Forests should be green with trees and plants. Oṣadhi-Sūkta of *Ṛgveda* addresses to plants and vegetables as mother, "O Mother! Hundreds are your birthplaces and thousands are your shoots".[56] The plants came to existence on their earth before the creation of animals.[57] *Chāndogya Upaniṣad* elaborates "water has generated plants which in turn generated food".[58]

52. *Ṛgveda*, 10.90.8.
53. *Yajurveda*, 19.20, 3.37; *Atharvaveda*, 11.2.24.
54. N.M. Kansara, 1995, *Agriculture and Animal Husbandry in the Vedas*, Delhi: Nag Publishers, pp. 126-38.
55. *Ṛgveda* 10.146.
56. *śataṁ vo amba dhāmāni sahasramuta vo ruhaḥ* — *Ṛgveda*, 10.97.2.
57. *yā oṣadhīḥ pūrvā jātā devebhyastriyugaṁ purā* — ibid., 10.97.1.
58. *tā annamasṛjanta. tasyādyatra kvacana varṣati tadeva bhūyiṣṭhamannaṁ bhavati* — *Chāndogya Up*. 6.2.4.

The *Atharvaveda* mentions certain names of *oṣadhis* with their values. Later this information became important source for the Āyurveda. The *Ṛgveda* instructs that forests should not be destroyed.[59] The *Atharvaveda* talks about the relation of plants with earth, "The earth is keeper of creation, container of forests, trees and herbs".[60] Plants are live.[61]

There is an important quotation in a Purāṇa which says, "One tree is equal to ten sons".[62] The *Atharvaveda* prays for continuous growth of herbs: "O Earth! What on you, I dig out, let that quickly grow over".[63] And another prayer says, "O Earth! Let me not hit your vitals".[64]

The *avi* element referred in the *Atharvaveda*, as the cause of greenness in trees,[65] is considered generally by Vedic scholars as "chlorophyll". The term *avi* is derived from the root *av* and thus gives the direct meaning of "protector". Hence, plants were studied as a part of environment and their protection was prescribed by the Vedic seers.

Concept of Sacrifice Yajña

The sacrifice *yajña* is regarded as an important concept of Vedic philosophy and religion but when we study it in its broader sense, it seems to be a part of Vedic environmental science.

---

59. *vanāni na prajahitāni* — *Ṛgveda*, 8.1.13.
60. *mandrāgretvarī bhuvanasya gopā vanaspatīnāṁ gṛbhiroṣadhīnām* — *Atharvaveda* 12.1.57.
61. *yena prāṇanti vīrudhāḥ* — ibid., 1.32.1.
    yathā vṛkṣo vanaspatistathaiva puruṣo mṛṣā — *Bṛhadāraṇyaka Up*. 3.9.28.
62. *daśaputrasamo drumaḥ* — *Padma Purāṇa* 1.44.45.
63. *yatte bhūmi vikhanāmi kṣipraṁ tadapi rohatu* — *Atharvaveda*, 12.1.35.
64. *mā te marma vimṛgvarī* — ibid., 12.1.35.
    pṛthivī mātarmā hiṁsī mā ahaṁ tvām — *Yajurveda* 10.23.
65. *avirvai nāma devatā . . . tasya rūpeṇeme vṛkṣāḥ haritāḥ haritasrajaḥ* — *Atharvaveda* 10.8.31; Kapil Dev, Dwivedi, *Vedic Sāhitya Evam Saṁskṛti*, Varanasi: Vishwavidyalaya Prakashan, 2000, p. 337.

Yajurveda and *Ṛgveda* describe it as the "navel (nucleus) of the whole world."[66] It hints that *yajña* is regarded as a source of nourishment and life for the world, just as navel is for the child.

Vedas speak highly of *yajña*. Through it, seers were able to understand the true meaning of the *mantras*.[67] All sorts of knowledge was created by *yajña*.[68] It is considered as the noblest action.[69] In simple words, *yajña* signify the theory of give and take. The sacrifice simply has three aspects: *dravya* (material), *devatā* (deity) and *dāna* (giving). When some material is offered to a deity with adoration, then it becomes *yajña*. Pleasing deity returns desired material in some different forms to the devotee. This *yajña* is going on in the universe since beginning of the creation and almost everywhere for production and also for keeping maintenance in the world. Even the creation of universe is explained as *yajña* in the Puruṣa-Sūkta. Thus, the concept of *yajña* seems to be a major principle of ancient environmental science.

In environment all elements are interrelated, and affect each other. Sun is drawing water from ocean through rays. Earth gets rain from sky and grows plants. Plants produce food for living beings. The whole process of nature is nothing but a sort of *yajña*. This is essential for maintenance of environmental constituents. The view that *yajña* cleans atmosphere through its medicinal smoke, and provides longevity, breath, vision, etc. is established in *Yajurveda*.[70] Few scholars have attempted to study the scientific nature of the

---

66. *ayaṁ yajño bhuvanasya nābhiḥ* — *Yajurveda*, 13.62; *ayaṁ yajño bhuvanasya nābhiḥ* — *Ṛgveda* 1.164.35.
67. *yajñena vācaḥ padavīyamāyamāyan* — *Ṛgveda* 10.71.3.
68. *tasmād yajñāt sarvahutaṁ ṛcaḥ sāmāni jajñire* — ibid., 10.90.9.
69. *yajño vai śreṣṭhatamaṁ karma* — *Śatapatha Brā*. 1.7.1.5.
70. *āyuryajñena kalpatām. prāṇo yajñena kalpatām. cakṣuryajñena kalpatām* — *Yajurveda* 9.21.

Vedic *yajñas*.[71] Undoubtedly, they have never been simple religious rituals, but have a very minute scientific foundation based on fundamental principles. According to Vedic thought, *yajña* is beneficial to both individual and the community. *Yajña* helps in minimising air pollution, in increasing crop yield, in protecting plants from diseases, as well as in providing a disease-free, pure and energised environment for all, offering peace and happiness of mind. Moreover, *yajña* serves as a bridge between desire and fulfilment.

Co-ordination Between all Natural Powers

Modern Indian scientists should be astonished and also feel proud of our ancestors for their knowledge and views about environment. Ancient seers knew about various aspects of environment, about cosmic order, and also about the importance of co-ordination between all natural powers for universal peace and harmony. When they pray for peace at all levels in the *śānti-mantra* they side by side express their belief about the importance of coordination and interrelationship among all natural powers and regions. The prayer says that not only regions, waters, plants, trees, natural energies but all creatures should live in harmony and peace. Peace should remain everywhere. The *mantra* talks about the concord with the universe:

> Peace of sky, peace of mid-region, peace of earth, peace of waters, peace of plants, peace of trees, peace of all-gods, peace of *Brahman*, peace of universe, peace of peace; May that peace come to me![72]

---

71. M.L. Gupta, 1999, *The Cosmic Yajña*, Jaipur: *Samhita Books* pp. 46-47.
72. *dyauḥ śāntirantarikṣaṁ śāntiḥ pr̥thivī śāntirūpaḥ śāntiroṣadhayaḥ śāntiḥ. vanaspatayaḥ śāntirviśve devāḥ śāntirbrahma śāntiḥ sarvaṁ śāntiḥ śāntireva śāntiḥ sā mā śāntirdhi* — *Yajurveda* 36.17; *Atharvaveda* 19.9.94; A.C. Bose, 1999, *The Call of the Vedas*, Mumbai: Bhartiya Vidya Bhavan, p. 281.

Conclusion

From the above detailed discussion some light is thrown on the awareness of our ancient seers about the environment, and its constituents. It is clear that the Vedic vision to live in harmony with environment was not merely physical but was far wider and much comprehensive. The Vedic people desired to live a life of 100 years[73] and this wish can be fulfilled only when environment will be unpolluted, clean and peaceful.

The knowledge of Vedic sciences is meant to save the human beings from falling into an utter darkness of ignorance. The unity in diversity is the message of Vedic physical and metaphysical sciences. Essence of the environmental studies in the Vedas can be put here by quoting a partial *mantra* of the *Īśāvāsyopaniṣad*: "One should enjoy with renouncing or giving up others part."[74] Vedic message is clear that environment belongs to all living beings, so it needs protection by all, for the welfare of all. Thus the study proves the origin of environmental studies from the Vedas.

---

73. *jīvema śaradaḥ śatam* — *Atharvaveda* 19.67.1.

74. *tena tyaktena bhuuñjīthāḥ* — *Īśāvāsyopaniṣad*, 1.

13

Contribution of Ancient India in Production Technology and Mechanical Engineering

Chandrabhan Prajapati

We have been told through Indian as well as foreign literary sources that in ancient times, commodities like sugar, palm oil, coconut oil, cotton cloth, clarified butter, cast iron, tin sheets, copper vessels, dyes and pigments like cinnabar (ochre), indigo and lac, perfumes like sandalwood oil, musk, tamarind, costus, macir, camphor, and even crude glass crockery were being exported from India. These items are not gifts of nature; their manufacture involves processing to effect chemical changes in their properties notably in the case of sugar, glass, metals and perfumes. Thus some kind of chemical engineering must have existed in India in those times, i.e. about 2000 to 2500 years ago. Along with this chemical processing, some physical apparatus would have been used. This presumes the existence of at least a rudimentary knowledge that in today's terminology would be called "mechanical engineering". The aim of this paper is to expose list of various commodities that were exported from India in ancient times. The present-day English names of most of these commodities have originated from Sanskrit. This list has been compiled from references to India made by Western, Chinese and Arab historians in ancient and medieval ages. In this paper we shall have a detailed look at some of these products which were imported from ancient India.

Introduction

Many of the advances in the sciences that we consider today

to have been made in Europe were in fact made in India centuries ago. — Grant Duff, British Historian of India

When referring to India, the author of the Greek text Periplus, which is dated around the first century CE, has said,

> There is a river near it called the Ganges. . . . On its bank is a market town which has the same name as the river, Ganges. Through this place are brought malabathrum and Gangetic spikenard and pearls and muslins of the finest sorts, which are called Gangetic. It is said that there are gold mines near these places, and there is a gold coin which is called caltis. And just opposite this river there is an island in the ocean, the last part of the inhabited world towards the east, under the rising sun itself, it is called Chryse; and it has the best tortoise-shell of all the places on the Erythrean Sea.

The Periplus further states that

> Nelcynda is distant from Muziris by river and sea about five hundred stadia, and is of another kingdom, the Pandian. This place also is situated on a river, about one hundred and twenty stadia from the sea. They send large ships to these market towns on account of the great quantity and bulk of pepper and malabathrum. There are imported here, in the first place, a great quantity of coin; topaz, thin clothing, . . . fine linen, antimony, coral, crude glass, copper, tin, lead; wine, not much, but as much as at Barygaza; realgar and orpiment; and wheat enough for sailors, . . . There is exported pepper which is produced in quantity only in one region near these markets, a district called Cottonara. Besides this there are exported great quantities of fine pearls, ivory, silk cloth, spikenard from the Ganges, malabathrum from the places in the interior, transparent stones of all kinds, diamonds and sapphires.

The Making of Dyes and Pigments

Many dyes and pigments were extracted in ancient India from

vegetable and mineral bases. The Greek historian Ktesias who lived in the fourth century BCE at the Persian court has observed that

> Among the Indians are found certain insects about the size of beetles and of a colour so red that at first sight one might mistake them for cinnabar. Their legs are of extraordinary length and soft to the touch. They grow upon trees which produce amber, and subsist upon their fruit. The Indians collect them for the sake of the purple dye, which they yield when crushed. This dye is used for tinting with purple not only their outer and undergarments, but also any other substance where a purple hue is required. Robes tinted with this purple are sent to the Persian king, for Indian purple is thought by the Persians to be marvellously beautiful and far superior to their own.

Ktesias also says that the Indian dye is deeper and more brilliant than the renowned Lydian Purple.

We will discuss some of these dyes which were extracted in ancient India, e.g. indigo, lac, ochre, copal, aniline, in the following headings.

LAC

Lac is a resinous substance secreted on trees by an insect called the lac insect. This is used as a protective covering and as varnish on wooden furniture. According to the *Oxford Dictionary*, the English word lac is derived from the Sanskrit term *lakṣa* which itself is derived from the word *rakṣa* which in Sanskrit means protection. This was perhaps as lac was used as a protective covering. Lac was also used as a dye. In ancient times it was used by women for dyeing nails and palms. It was also used to dye cloth. The process of dyeing cloth with lac was termed *vastra-raṅga-kṛta* which literally means "to give colour to cloth." In doing this screens were used to hold cloth in place while the dye was being applied.

In ancient times, lac was used both as a dye and a resin but with the introduction of aniline, the demand for lac as a dye became less. It is today used mainly as a resin called "shellac", which is melted into thin flakes and used to make varnish.

OCHRE

Ochre is a pigment varying from light yellow to orange and brown. It is a mineral of clay and hydrated ferric oxide. The old name of ochre is cinnabar which was perhaps derived from the Sanskrit term *sindūra*. According to the *Twentieth Century Dictionary*, the word "cinnabar" originated from the Persian word *zinjafr*, which is a corruption of the Sanskrit *sindūra*.

Ochre (or cinnabar) was used as a dye for cloth and also as paint for walls in Roman times. In India it was, and still is used to paint images of gods and as a caste mark applied on the forehead called *tilaka*.

COPAL

Copal is a resin extracted from a tree which in Latin is named *Vateria Indica*. This tree is to be found in the Western Ghats (Hills) of India. The word "copal" is not to be found today in the *Oxford Dictionary*, but according to the Periplus, "it is derived from the Sanskrit term *kaṅkamon*.

ANILINE

This is a blue-coloured dye obtained from coal tar. This was being extracted in ancient India and was transmitted to the West by the Arabs. It was called *nīla* or *nīlama* in Sanskrit. The Arabs named it *ai nil* or *an nil* from which we have the English word aniline. This is corroborated by the *Oxford Dictionary*.

INDIGO

Indigo is also a blue-coloured dye obtained from a plant named

Indigofera. In ancient times, indigo was used both as a dye and as a medicine. The word "indigo" is derived from the Greek word *Indikon* which means "from India". In Sanskrit it is referred to as *nīlama*.

The Smelting of Metals

According to information culled out from various Roman and Greek texts, metals like iron, tin, copper and brass were imported from India. These texts say that the metals were not being imported as an ore but as sheets. This presumes that the ore must have been smelted and cast into sheets in India before it was exported. References in Sanskrit literature also support this.

According to Ktesias (Greek author living in Persia in fourth century BCE), the smelting of metals was undertaken in India in those days. He has written that,

> Every year a spring filled itself with fluid gold which was drawn from it in one hundred earthen pitchers. It was necessary that they should be of clay, because the gold afterwards congealed, and the pitchers had to be broken in order to get it out. . . . Each pitcher contained one talent of gold.

On this remark of Ktesias, McCrindle, who has translated Ktesias' writings, has noted that

> The sense of this passage can only be that auriferous ores were melted, and that the gold obtained from them was drawn out in a fluid state. That there was a spring must be a misapprehension, and we must imagine instead that there was a cistern prepared to receive gold. . . . If this supposition is right, it follows that the Indians knew how to extract gold from the ore by melting.

We shall now look into some of the metals that were being exported from India.

IRON

Marco Polo has mentioned that iron and Ondanique was sold in the markets of Kerman in Iran. The word *Ondanique* has been interpreted as a corruption of the Persian word *Hindwaniy* which meant "Indian Steel".

Even earlier, during the reigns of the Roman emperors Marcus Aurelius and Commodus, Ferrum Indium appears in the list of dutiable articles. There also exists an ancient Greek chemical treatise entitled *On the Tempering of Indian Steel*. Edrisi has noted that

> The Hindus excel in the manufacture of iron. They have also workshops wherein are forged the most famous sabres in the world. It is impossible to find anything to surpass the edge that you get from Indian Steel.

This passage which has been quoted in the notes to the Periplus on page 71 proves beyond doubt, in the words of a foreign historian, that the art of smelting and casting iron was well developed in ancient India.

In ancient times, in India, *loha-cūrṇa* meant iron ore; *kupya-śālā* and *sandhāni* meant an iron foundry. A furnace was called *cūlī* or *agnikuṇḍa*. Wrought iron was called *lohabandhana*; iron bars were called *loha-piṇḍaha*. Smelting of iron was called *loha-drāva-karaṇa* (literally, liquefaction of Iron). *Loha-cihna* meant an iron mould and *lohakāraka* meant a smith or ironmonger.

TIN

In ancient tunes tin was known as Kassiteros in the Greek language. This word was derived from the early Sanskrit word *kasthira* for tin. Even today in some Indian languages tin and aluminium are called *kaṭhila* which is derived from the Sanskrit word *kasthira*.

In ancient India, the value of tin for hardening copper was recognised and the art of tempering tin with other metals was developed. The fact is corroborated by the comments of the Greek historians regarding the excellent tempering of various metals that had been practised in India.

COPPER

Copper has been mentioned in the Periplus as an article of export from India. In those days copper ore was extracted in a big way and it was smelted locally in south India and Rājpūtānā, according to the Periplus.

Philostratus of Lemnos, in about CE 230, has mentioned a shrine in Taxila in India, in which were hung pictures on copper tablets representing the feats of Alexander and Porus. In the words of Philostratus

> The various figures were portrayed in a mosaic of orichalcum, silver, gold, and oxidised copper, but the weapons in iron. The metals were so ingeniously worked into one another that the pictures which they formed were comparable to the productions of the most famous Greek artists.

In ancient India, copper was also known as *tāmra,* copperplate was called *tāmra-patra. Tāmrakāra* meant a coppersmith and *tāmra-paṇa* meant a copper coin.

BRASS

Brass is not an original metal obtained from natural ore, it is a composite metal found as an alloy of copper and zinc. From the above quotation of Philostratus it is evident that in ancient India the art of welding metals together was known. We cannot conclusively say that brass (called *pītalam* in Sanskrit) was manufactured in ancient India but the fact that there were Sanskrit equivalents for copper (*tāmra, kasthira*), zinc (*dastā*) an alloy which was called *miśradhātu* (mixed metal) *nyūnatā* (novelty) and *dūṣatā* (impurity) supports the fact that brass, or a brass-like metal, was smelted in ancient India.

Thus it was recognised that an alloy would be a mixture of metals, that it would be novelty; and that due to the mixing of metals, the resultant alloy would be impure in the sense that it would not be a metal derived from a natural mineral ore.

It is possible that brass was used for decorative purposes due to its likeness to gold. Thus we have the terms like *pītalam-puśpakam, pītalam kusuman janam* and *pītalam-puśpaketu* which mean efflorescence created from brass. The ornamental use of brass in ancient India is also evident from the quotation of Philostratus given above in the section on copper.

CORUNDUM

This unfamiliar word connotes crystallised aluminium oxide of various colours having great hardness and used as gems and also as an abrasive. We have been told that certain gems were used to cut metals in ancient India. Corundum or *kuruvinda* was one such gem. Corundum was also exported from India to foreign countries since ancient times. The fact that India was the main source of this item is evident from the name Corundum given to it which according to the *Oxford Dictionary* is derived from the Sanskrit word *kuruvinda*.

NAUŚĀDAR

This is another unfamiliar term which means chloride of ammonium. This was also being obtained from ancient India. The word *nauśādar* is also derived from a Sanskrit root word *narasāra*.

BERYL

This is a mineral species of a transparent precious stone varying from pale green to yellow in colour. This was one of the items of export from India during early times. The word Beryl is said to have been derived from the Sanskrit term *vaidūrya*.

COPAL

This is amorphous quartz-like form of hydrated silica; some types of opal are semi-translucent and appear like glass. This item is recorded to have been exported in ancient times from India. The English word "opal" has been derived from the Sanskrit term *upala* a fact which is corroborated by the *Oxford Dictionary*.

Conclusion

From this survey of the different products that were being manufactured and exported we have seen that since ancient times India had developed the mechanical arts. Metals were smelted and cast into sheets, even into pictures as was recorded by Greek historians; fabric was woven and dyed, needless to say yarn was spun to produce the finest muslin fabrics; oil was extracted from coconuts, groundnuts, palm and sesame; perfumes were made from sandalwood, musk, camphor and spikenard; dyes and pigments were extracted from ochre, indigo, lac, and copal; sugar crystals were made, coir mats woven and even glass and imitation jewellery was manufactured.

Obviously, all this could not have been done with bare hands. The techniques used in making sugar, glass, jewellery, metal sheets, oils, etc. presumed the existence of physical apparatus of some degree of sophistication.

To fabricate that physical apparatus knowledge of mechanics and mechanical engineering was a prerequisite. The various references to the use of spindles, oil-mills, furnaces given under various sections in this paper throw light as to the state of mechanical arts in ancient India which supported our export trade and enabled us to maintain a monopoly in the supply of many items for centuries together.

Endnotes

1. W. Caland (ed.), 1931/1982, *Pañcaviṁśa-Brāhmaṇa: The Brāhmaṇa of Twenty Five Chapters*, Calcutta: Asiatic Society.
2. M.R. Bhat, 1995, *Varāhamihira's Bṛhat Saṁhitā*, Delhi: Motilal Banarsidass.
3. R. Kaulacāra, 1966, *Śilpa Prakāśa*, ed. A. Boner, and S. Rath Sarma, Leiden: E.J. Brill.
4. S. Kak, 2003, "Early Indian Music", in *A Search in Asia for a New Theory of Music.*
5. J.S. Buenconsejo (ed.), University of the Philippines, Center for Ethnomusicology Publication, Quezon City; http://www.ece.lsu.edu/kak/manila.pdf
6. B.B. Lal, 1997, *The Earliest Civilization of South Asia*, New Delhi: Aryan Books International.

Part - V
Yoga and Āyurveda

14

Personality Traits of People Born in Different Saṁvatsaras and their Comparison with Modern Psychological Personality Traits
A Study

Melukote K. Sridhar

Astrology and later astronomy have had a profound influence over the past few millennia on both Eastern and Western cultures. It was believed that heavenly bodies had a direct or indirect influence upon people, their lives from birth to death and even beyond. Both uneducated and educated people were keen to know their future in all dimensions such as way of life, achievements, economic conditions, health, long life, love, marriage, offsprings, calamity, etc., and hence they resorted to this branch of learning, irrespective of criticism from rationalists, non-believers, intellectuals and the like.

Indian traditional astronomers such as Bhṛgu, Puṇḍit Dhuṇḍi Rāja, Varāhamihira in their astronomical treatises and compilations (Saṁhitās) have spoken extensively about the effects of *saṁvatsaras* (one full year) on individuals and the world at large. *Saṁvatsara* is a time in which planet Jupiter (*guru gṛha*) crosses the sign of one zodiacal sign (*rāśi*) with his average speed and will be nearest or farthest to the earth and is in touch with auspicious or inauspicious rays. It takes sixty years for Jupiter and Saturn (*śani*) to reach the zero degree of the Aries (*meṣa*) sign. Various angles found between Jupiter and Saturn have their influence on *saṁvatsaras*.

The traditional astrologers say that a person born in each *saṁvatsara*, month (*māsa*), fortnight (*pakṣa*), lunar day (*tithi*) and day (*dina*) has different personality traits and characteristics compared to other members of the same family. The modern cardiologists such as Meyer Friedman and R.H. Rosenman (1950) have compared the four types of personality such as A, B, C, and D and a combination of A-B, C-D, etc. for knowing health patterns. These personality types exhibit same type of behavioural patterns worldwide and have striking similarities with the people born in different *saṁvatsaras*.

In this paper, I propose to enlist the characteristics of different *saṁvatsaras* (from *prabhava* to *krodhana*), the behaviour and characteristics of people born in such *saṁvatsaras* and compare them with the modern personality types. A set of horoscopes would be analysed as a part of this study. It is hoped that such an inter-disciplinary study covering the Indian knowledge system with the modern knowledge system would reveal the ancient Indian wisdom in a new perspective for the benefit of mankind.

Keywords: Astrology, Day, Horoscope, Jupiter, Personality types, *saṁvatsara*, Saturn.

Preamble

ASTROLOGY had a profound influence over the past few millennia on both Eastern and Western cultures. In the medieval ages the educated people believed in astrology, the system of heavenly spheres and bodies which were believed to reflect on the system of knowledge and the world below. Belief in astrology holds firm today in many parts of the world: in a poll conducted by an NGO in the West, 31 per cent of Americans expressed a belief in astrology and, according to another study, 39 per cent considered it as scientific.[1]

Astrology had a terrific influence on both language and

1. (a) Humphrey Taylor, "The Religious and Other Beliefs of Americans 2003", http://www.harrisinteractive.com/harris_poll/index.asp?PID=359. Retrieved 5-1-2007.

   (b) "Science and Technology: Public Attitudes and Understanding", National Science Foundation. http://www.nsf.gov/statistics/seind06/c7/c7s2.htm#c7s213. Retrieved 5-1-2007.

literature. For instance, influenza, from medieval Latin *influentia* meaning influence, was so named because doctors once believed that this epidemic was caused by unfavourable planetary and stellar influences.[2] The word "disaster" comes from the Italian *disastro*, derived from the negative prefix *dis-* and from Latin *aster* "star", thus meaning "ill-starred"[3] Adjectives "lunatic" (Luna/Moon), "mercurial" (Mercury), "venereal" (Venus), "martial" (Mars), "jovial" (Jupiter/Jove), and "saturnine" (Saturn) are all old words used to describe personal qualities said to resemble or be highly influenced by the astrological characteristics of the planet, some of which were derived from the attributes of the ancient Roman gods they are named after. In literature, many writers, notably Geoffrey Chaucer[4] and William Shakespeare,[5] used astrological symbolism to add subtlety and nuance to the description of their characters' motivation(s). More recently, Michael Ward has proposed that C.S. Lewis imbued his *Chronicles of Narnia* with the characteristics and symbols of the seven heavens. Often, an understanding of astrological symbolism is needed to fully appreciate such literature.

---

2. http://www.etymonline.com/index.php?term=influenza Online Etymology Dictionary.
3. "Online Etymology Dictionary: Disaster", http://www.etymonline.com/index.php?term=disaster. Retrieved 22-1-2009.
4. (a) A. Kitson, "Astrology and English literature", *Contemporary Review*, October 1996, http://www.findarticles.com/p/articles/mi_m2242/is_n1569_v269/ai_18920172. Retrieved 17-7-2006.

   (b) M. Allen, J.H. Fisher, "Essential Chaucer: Science, Including Astrology". University of Texas, San Antonio. http://colfa.utsa.edu/chaucer/ec22.html. Retrieved 17-7-2006.

   (c) A.B.P. Mattar et al., "Astronomy and Astrology in the Works of Chaucer". University of Singapore.http://www.math.nus.edu.sg aslaksen/gem-projects/hm/astronomy and astrology in the works of chaucer.pdf. Retrieved 17-7-2006.
5. (a) P. Brown, "Shakespeare, Astrology, and Alchemy: A Critical and Historical Perspective", The Mountain Astrologer, February/March 2004. http://www.astrofuturetrends.com/id19.html.

   (b) F. Piechoski, "Shakespeare's Astrology", http://starcats.com/anima/shakespeare.html.

Some modern thinkers, notably Carl Jung,[6] believed in astrology's descriptive powers regarding the mind without necessarily subscribing to its predictive claims. In education astrology is reflected in the university education of medieval Europe, which was divided into seven distinct areas, each represented by a particular planet and known as the seven liberal arts. Dante Alighieri speculated that these arts, which grew into the sciences we know today, fitted the same structure as the planets. In music the best known example of astrology's influence is in the orchestral suite called "The Planets" by the British composer Gustav Holst, the framework of which is based upon the astrological symbolism of the planets.

Saṁvatsaras and their Significance

In the Indian context and scenario, something connected to life in general may be happening in the sixty-year *saṁvatsara*. Etymologically *saṁvatsara* is derived from the root of the Sanskrit verb *vas* meaning *to dwell* with preposition *saṁ* meaning *together with* and the word *sara* meaning *moving*. Thus the word denotes the movement of six seasons, or one year and the dwelling of people in that year (*saṁvasanti rutavaha atra – saṁ + vas (nivase) + saraha* (*Uṇādi* III.72). But the names of *saṁvatsara* puzzle us — *prabhava* through *kṣaya*. They depict birth, growth, decay, and death of a person (15+15+15+15). So, something connected to life in general may be happening in the sixty-year *saṁvatsara* cycle.[7]

Paṇḍit Dhuṇḍi Rāja, the author of *Jātakabharaṇam*, having studied the astrological compilation texts (*saṁhitā grantha*) has

6. Carl G. Jung, "Archetypes of the Collective Unconscious", excerpted in *The Basic Writings of C.G. Jung* (Modern Library, repr. 1993), pp. 362-63.
7. (a) Roger Beck, *A Brief History of Ancient Astrology*, Blackwell, USA (2007).

   (b) Paramahansa Yogananda, *Autobiography of a Yogi*, Chapter 16: "Outwitting the Stars".

spoken of the effects of every *saṁvatsara* on an individual, while their effects have greater importance in the context of all people depending upon the time and place of birth. By generalising the effects of the *saṁvatsara* (year), month (*māsa*), fortnight (*pakṣa*), lunar date (*tithi*) and day (*dina*), much knowledge may be gained about the inclination or tendency, health, education, profession, achievements, economic conditions, love and married life, longevity, success and failures in life, etc. of every individual.

Saṁvatsara generally means the period of one full year. The time in which the planet Jupiter (*guru*) with his average speed crosses the journey of one zodiacal sign (*rāśi*) is referred to as *saṁvatsara*. There occur twelve *saṁvatsara*s when Jupiter completes one complete round of the zodiac. While moving in the zodiac, Jupiter happens to reach sometimes near the earth and sometimes far away from it. Besides, it also comes in contact with auspicious or inauspicious rays during its journey. It is because of this fact that the effects of every *saṁvatsara* become different. Sixty *saṁvatsara*s repeat again and again in a cycle. At the beginning of creation Jupiter and Saturn were on the zero degree of Arics (*meṣa*). The average daily motion of Saturn is two minutes (*kāla*s). Therefore, it takes full 30 years in revolving around the zodiac. The average daily motion of Jupiter is five minutes (*kāla*s). Therefore it takes full 12 years in completing its journey in the twelve signs of the zodiac. The L.C.M. (Lowest Common Multiple) of 30 and 12 is 60. Therefore, at every 60 years, both the planets (i.e. Jupiter and Saturn) reach the same zero degree of the sign of Aries. In this journey, the various angles that are formed between Jupiter and Saturn have their influence on the effects of the *saṁvatsara*s. As a matter of fact, no other relationship of the other planets affects an individual, a place or a country so much as the mutual angular relationship of Saturn and Jupiter. During his journey Jupiter (*guru*) contacts auspicious or inauspicious rays in the galaxy, stars, planets, etc. The movements of Jupiter and Saturn in 2009-10 (virodhi *saṁvatsara*) were as follows: Jupiter (*guru*) moved in the zodiacal sign of Capricorn (*makara rāśi*) from 27-03-2009 to 1-5-2009, 30-07-2009 to 19-12-2009 and in Aquarius (*kumbha rāśī*) from 1-5-2009 to 15-03-2010. Saturn (*śani*) moved in Leo (*siṁha rāśī*) from 1-5-2009 to 09-09-2009 and in Virgo (*kanyā rāśi*) from 10-09-2009 to 15-03-2010. In *vikṛti saṁvatsara* (2010-11) Jupiter moved in the zodiacal sign of Aquarius (*kumbha rāśi*) from 16-03-2010 to 2-5-2010 and again from 1-11-2010 to 6-12-2010. Further, Jupiter moved in the zodiacal sign of Pisces (*mīna rāśi*) from 2-5-2010 to 1-11-2010 and from 6-12-2010 to 3-11-2011. Saturn (*śani*) moved in the zodiacal sign of Libra (*tulā rāśi*) throughout the year (16-3-2010 to 3-11-2011).

Major delineations of *Bhṛgu Saṁhitā* seem to have been based on the relationship of these two planets, i.e. Jupiter and Saturn. In the books on *Siddhānta Jyotiṣa* (Astronomy) the *saṁvatsara*s have been counted 60 beginning from the *saṁvatsara* named Vijaya and ending with Nandana (the 60th *saṁvatsara*). But Varāhamihira in *Bṛhat Saṁhitā* has regarded the beginning of the *saṁvatsara* cycle, not from the Vijaya, but from the 35th *saṁvatsara* of *prabhava*. Since Varāhamihira's times, it is this sequence which has been accepted by astronomers and astrologers all over the country.[8] The 60 *saṁvatsaras* and their general characteristics on people as explained in traditional astrological texts are as follows.

Saṁvatsaras: General Characteristics

1. *Prabhava* — Readiness for the collection of all kinds of things, blessed with sons, has excellent intellect, enjoys all kinds of comforts, longevity.

---

8. Summary translation of an article titled *Effects of Samvatsars*, 18, September 2008, Posted by Neelam on the Internet.

2. *Vibhava* — Enjoys things which are meant to be consumed (foods, drinks, etc.), extremely beautiful, strong and intelligent, knows the mysteries of arts, good mannered, cultured and very learned.
3. *Śukla* — Remains joyful, extremely generous, has excellent qualities, blessed with sons, full of grandeur and prosperity, endowed with good fortune, learning and humility.
4. *Pramoda* — Donor, blessed with sons, enjoys pleasure, extremely beautiful, truthful, has good qualities, skillful, deceptive, noble and proud.
5. *Prajāpati* — Proud of a wife endowed with good qualities, kind-hearted, practises religion of his family, has excellent nature, does reverence to gods, elders, teachers, and courteous.
6. *Aṅgirasa* — Endowed with beauty, happy, self-pride, sweet-spoken and blessed with sons, trustworthy, long-lived.
7. *Śrīmukha* — Wealthy, valorous, energetic, knowledgeable, affectionate, truthful, intelligent, physically strong, extremely generous.
8. *Bhāva* — Has splendid or magnificent mind, very famous, munificent, courteous, happy, loved by all.
9. *Yuva* — Direct incarnation of happiness, has good qualities, courteous, peaceful, generous, full of erudition or learning, long-lived, has robust body, contented.
10. *Dhātrī* — Has good qualities, beautiful, devoted to teacher, skilled, courteous and good-mannered.
11. *Bahudhānya* — Wise in trade, respected, charitable, suffers from pride, scholarly, and wealthy.
12. *Īśvara* — Short-tempered, full of joy, has good qualities,

valorous, skilful, sagacious, skilled in fine arts and courteous.

13. *Pramathi* — Owner of vehicles, flag, umbrella, horses, studies treatises, enemy-killer, minister of kings, has knowledge of ancient texts.
14. *Vikrama* — Engaged in doing extremely terrible or fierce deeds, skilled in attacking enemy's army, warrior, has patience and endurance, extremely generous, valorous and powerful.
15. *Vṛṣa* — Praises work done by his own self, blameworthy, has company of vicious people, helps others, and has many wives, dirty, lazy and greedy.
16. *Citrabhānu* — Fond of various kinds of clothes and flowers, ambitious, good natured and endowed with crafts or arts.
17. *Svarbhānu* — Possesses curly hair, simple nature, beautiful, conqueror of enemies, wise, courteous, smiling, endowed with splendour or magnificence.
18. *Tāraṇa* — Cunning, heroic, restless, extremely harsh, cruel, engages in wicked acts, consumer and is endowed with wealth.
19. *Pārthiva* — Practises religion, skilled in arts, lover of sensual pleasures, dominant.
20. *Vyaya* — Engrossed in enjoying worldly pleasures, subjected to addictions, non-hesitant in asking for debts, always remains in debt, restless, has a tendency of spending much.
21. *Sarvajit* — Honoured by the king, pure, has a huge body, conqueror of the king as well as his enemies.
22. *Sarvadhārī* — Master of a retinue of servants, enjoys pleasures, endowed with beauty, likes sweets, has

patience, endurance and obeys traditions and conventions.

23. *Virodhī* (2009-10) — Eloquent speaker, wanders in foreign lands, rarely gives joy and happiness to his own people, extremely deceitful, enmity with people.
24. *Vikṛti* (2010-11) — Subjected to poverty, horrible-looking, has tall body, is given to pride and is lacking in wisdom and intelligence, has less friends.
25. *Khara* (2011-12) — Lustful, dirty in his body, speaker of very harsh and loud words without any reason or purpose, quarrelling, shameless and possesses a huge body.
26. *Nandana* — Interested in donating grains as alms, has pure heart and remains happy with good wife and children.
27. *Vijaya* — Brave, has good conduct, honoured by kings, an excellent speaker, bountiful, kind-hearted and enemy killer.
28. *Jaya* — Organiser of philosophical deliberations, honoured in the world, bountiful and generous, destroyer of enemies, longing desire for getting victory, engrossed in worldly or sensual enjoyment, very resplendent.
29. *Manmatha* — Adorned with ornament of special kind, sensuous, sweet-spoken, always remains engaged in singing and dancing and is voluptuous.
30. *Durmukhī* — Cruel, untrustworthy, endowed with reprehensible or blameworthy mentality, avaricious, has curved mouth, curved hands and feet, remains engrossed in sins, is extremely villainous and wicked.
31. *Hevilambī* — Endowed with horses of high speed and four-wheeled vehicles, gold, clothes, wealth and grains and gems, happy.

32. *Vilambī* — Deceitful, extremely greedy, lazy, phlegmatic, weak, fatalist and has the habit of speaking without purpose, serviceable, defeated by his wife, contented, keeps his thoughts secret, restless in nature.

33. *Vikārī* — Stubborn, skilled in arts, restless, cunning, extrovert and insecure.

34. *Śārvarī* — Skilled in trade activities, sensual, unfriendly, engaged in acquiring knowledge.

35. *Plava* — Extremely voluptuous, wealthy, gets respect, defeated by his wife, contented, keeps his thoughts secret, restless in nature.

36. *Śubhakṛta* — Endowed with good fortune, learning, courtesy or humility and excellent virtuous deeds, long-lived and have many sons, much wealth and property.

37. *Śobhākṛta* — Makes progress in every field, handsome-looking, has excellent virtues, kind-hearted, skillful, doer of good deeds, successful, brilliant, courteous, has beautiful eyes.

38. *Krodhī* — Cruel-eyed, cruel-natured, has great love for his wife, extremely haughty or proud, puts obstacles for others, and prone to anger.

39. *Viśvāvasu* — Blessed with virtuous spouse and son, excessively generous, has excellent conduct, excessive patience and endurance, likes sweet foods and endowed with all the virtues.

40. *Parābhava* — Can hardly make accumulation of wealth, speaker of bitter or harsh words, devoid of good conduct and is stupid.

41. *Plavaṅga* — Volatile or of restless mind, does not have a desire to do good deeds, deceitful, devoid of good conduct, thoughtless and has weak body.

42. *Kīlaka* — Average handsomeness, sweet-spoken, kind-hearted, has love for water, has very fat legs, beautiful forehead, strong and destroyer of his enemies.

43. *Saumya* — Destined to be a scholar, wealthy, very sensual, has love for his deity and guests, pure, endowed with noble habits, has weak body, reputation through the medium of trade.

44. *Sādhāraṇa* — Loves travelling, talented in writing, has discrimination, given to anger, pure and detached.

45. *Virodhīkṛta* — Remains engrossed in the worship of Lord Śiva, angry nature, quarrelsome, neglects his father.

46. *Paridhāvī* — Learned, courteous, adroit in the field of arts, wise, honoured in king's court.

47. *Pramādica* — Villainous or wicked, proud, quarrelsome, avaricious or greedy, has great love for his own men, destined to be poor, dumb and doer of deeds which are apprehensible or blameworthy.

48. *Ānanda* — Skilful, excessively adroit, endowed with happiness, have sons, learned, has sense of gratitude or thankfulness, courteous and liberal.

49. *Rākṣasa* — Extremely cruel, performer of reprehensible deeds, quarrelsome, devoid of religion and thoughtfulness, cruel and courageous.

50. *Anala* — Gifted with good sense, expert in the trade of things produced in water, good character, a little wealthy, restless, and supporter of many.

51. *Piṅgala* — Performer of reprehensible deeds, fierce nature, restless, has grandeur, beneficent, stupid and harsh speaking, remains happy, gets fame in the battle, handsome, minister of kings, honoured by many and capable or competent.

52. *Kālayukti* — Derives pleasure from speaking too much without any purpose, endowed with blame (base) intellect, devoid of fortune, Yama (the god of Death) incarnate when a quarrel arises and is weak-bodied.

53. *Siddhārthī* — Generous, remains happy, gets fame in the battle, handsome, minister of kings, honoured by many.

54. *Raudra* — Has bad appearance, rears cattle, speaks ill of others, excessively deceitful, gets a bad name, vicious heart and very fierce.

55. *Durmati* — Pride of keeping his words, remains constantly unhappy, sensuous, engaged in doing base deeds and foolish.

56. *Dundubhi* — Recipient of honours from kings, endowed with elephants, horses, lands and gold, lover of dance and songs.

57. *Rudhirodgārī* — Red-eyed, weak-bodied, excessively hot-tempered or, has bad nails on his hands and feet and gets wounded by some weapon.

58. *Raktākṣī* — Engages in righteous and religious conducts, extremely lustful, does not tolerate the growth of others.

59. *Krodhana* — Puts obstacles in the work of others, dominated in temperament by the quality of the three cardinal elements in man's nature, fierce or terrible and puts others in delusion.

60. *Akṣaya* — Spends the wealth earned by him very quickly, has the tendency of serving others, hard-hearted and does not have much desire to do good deeds.

3. Days (Dina): Significance

The traditional astrologers and astronomers have said that apart from the characteristics of a *saṁvatsara* in which a person is born, the nature of a person born on a specific day also will have distinctive characteristics as follows:

1. Sunday (*Bhānuvāsaraḥ*) — strong soul, powerful, fiery, dynamic, concern for parents.
2. Monday (*Induvāsaraḥ*) — fickle, intelligent, pleasant, concern for mother.
3. Tuesday (*Bhaumavāsaraḥ*) — acquires brothers, servants, land and other immovable property.
4. Wednesday (*Saumyavāsaraḥ*) — intelligent, courageous, has scientific and philosophical pursuits.
5. Thursday (*Guruvāsaraḥ*) — robust health, has leadership qualities, wisdom, male offsprings, earns profuse wealth.
6. Friday (*Śukravāsaraḥ*) — endowed with fortune, marriage, multiple partners, vehicles, etc.
7. Saturday (*Śanivāsaraḥ*) — has longevity, makes immoveable property, receives both major honour/insult.

According to the Indian lunar calendar (*Candramana Pañcāṅga*), the 30-day cycle is divided into two fortnights called the *śukla-pakṣa* and *kṛṣṇa-pakṣa* depending upon the waxing and waning of the moon respectively. The fifteenth day of *śukla-pakṣa* ends with Full Moon day (*Pūrṇimā*) while the fifteenth day of *kṛṣṇa-pakṣa* ends with New Moon day (*Amāvasyā*). Each fortnight (*pakṣa*) and the specific day in that fortnight has got its own significance. Generally it is observed that those born in the *śukla-pakṣa* are very active, dynamic and extroverts whereas those born in *kṛṣṇa-pakṣa* are mild in activity and are introverts by nature.

Modern Psychological Traits

The modern cardiologists such as Meyer Friedman and R.H. Rosenman (1950) have come out with four types of personality such as A, B, C, and D and a combination of A-B, C-D, etc.

PERSONALITY TYPE A

People with Personality Type A are generally impatient, time-conscious (cause irritation and exasperation), status- conscious, highly competitive, ambitious, business-like, aggressive, having difficulty in relaxing, low self-esteem and exhibit hostility towards others. They are highly independent and driven personality — leaders in business. Their ambition of leaving their mark in the world singles them out from the rest. They are self-driven and know the importance of goal setting, positive thinking and motivation. For instance the MD, CEOs of business establishments. They are blunt, competitive, no-nonsense types who like to get to the point. They have entrepreneurial spirits and take risks in all endeavours. They embrace change; look for practical solutions for solving problems. They are prone to competitiveness and have strong achievement orientation.[9] For instance they easily get frustrated while waiting in line, interrupt others, walk or talk at a rapid pace, are always painfully aware of time. They have the ability to put on the veil of practicality for solving a problem immediately. They are open-minded individuals who welcome changes in technology and practices for attaining success.

PERSONALITY TYPE B

People with Type B personality are high-achieving workaholics who engage in multi-tasks, drive themselves with deadlines, and are unhappy about delays. They are patient, relaxed, and easy-going, generally lacking any sense of urgency, apathetic and disengaged.

Such people are highly extroverts, are always very entertaining and possess strong charisma. Other people love to hang around them. They are human magnets that can attract

9. Meyer Friedman and R.H. Rosenman, *Type A Behavior and Your Heart*, New York, 1950.

the attention of the rest in a gathering, effortlessly. This makes their type seem special among the various other personalities. These individuals have a very engaging personality and when it comes to conversations are good marketing types. They thrive on the interaction with the rest. They easily get hurt and cannot tolerate comments or criticisms. They exhibit facial tension with tightened lips, clenched jaws, shivering body at the time of expressing anger. They have dark circles below their eyes and suffer from loss of good sleep. Such people suffer from hypertension, high blood pressure, heart disease, job stress and social isolation.

PERSONALITY TYPE C

People with Type C personality are antithesis of Type B. They are introverts, eager to know the precious details of people and things. These are introverts who are interested in details which separate them from the rest of the types in personality types A, B, C. They could turn heaven and hell upside down or inside-out, depending on their mood, to find the detail or fact they are interested in, before taking it into pieces in order to restart the process. They have trouble in communicating to people, are very cautious, reserved, seek pleasure, shirk away from social or human interactions. They rarely take risks. Arranging around them, in logical order is what drives these people to no end. This type of a person can live with an individual of personality Type B. Personality Type C is as exactly opposite of the personality Type B as an electron with the left spin is the antithesis of the one with the right spin. You can find specimens of personality Type C personified in accountants, computer programmers, etc. Reserved nature is a mark of personality Type C and they are cautious too. Risk taking is not a very attractive option for these people.

PERSONALITY TYPE D

These gentlemen really believe in inertia which they have no shortage of. These people prefer to stick to the trodden paths and established routines over the uncertainty of change. These are followers of the spent actions and executors of the direct commands. You will find them doing their best not to stretch their neck out when it comes to taking responsibility and risk. With the help of professionals in the field of personality development and through sheer power of self-motivation or self-improvement, these individuals can overcome their handicap to some extent.

Cardiologists Meyer Friedman and R.H. Rosenman described Personality Types A and B after a nine-year study of 9000 healthy people (35-59 years). They opined that Type A behaviour doubles the risk of coronary heart disease. This research had tremendous effect in stimulating the development in the field of health psychology, in which psychologists look at how a person's mental state affects a person's health.[10] It is not uncommon to find people with a blend of personalities, particularly A-B and C-D. But these basic personality types explain why some people work well together and others do not. For example, Type A clashes with Type D simply because one is more adventurous than the other and Type B clashes with Type C as one exhibits an extroverted personality and the other is introverted. Conversely Type A works well with Type B, and Type C works well with Type D.[11]

Analyses

- An analysis of 53 horoscopes (table given below) belonging to different *saṁvatsaras* and different days denote the following: Generally people born on

---

10. M. Friedman, *Type A Behaviour: Its Diagnosis and Treatment*, New York: Plenum Press (Kluwer Academic Press), 1996.
11. Tim Bryce, *Morphing into the Real World: A Handbook for Entering the Work Force*, MBA, Florida, 2007.

Sundays have dynamic/aggressive nature and are extroverts (Type A).

- Generally people born on Mondays are fickle in nature, pursue love matters, sometime eccentric, and have suffered from respiratory diseases (Type C).
- People born on Tuesdays are also aggressive in nature, have a tendency to do unholy things and quarrelsome (Type A-B admixture). People born on Wednesdays are sharp, intelligent, clever, intellectuals (Type B).
- People born on Thursdays are leaders in respective professions, have broad mind, people loved by all (Type A).
- People born on Fridays have excelled in fine arts, secretive in nature and engage in gallant actions (combination of Types A-B).
- People born on Saturdays are lazy in learning and work, introverts, very secretive in nature (Type C and D).
- People born on New Moon days have criminal outlook, are hyperactive during nights, amass lot of wealth and property, always restless and secretive in nature.
- People born on Full Moon days are quiet and composed, tend to do noble acts and receive social recognition.

Acknowledgement

I offer my grateful thanks to Prof. Bal Ram Singh, Director, Center for India Studies, University of Massachusetts, Dartmouth, USA, for giving me an opportunity to present my views at a Brainstorming Session at Hardwar (18-6-2009) and paper at the International Seminar titled "Provable Concepts of Science & Technology in Indian Texts (STAIT)" at Jawaharlal Nehru University, New Delhi, 9-1-2010, and to Dr Girish Nath Jha, Associate Professor, Center for Special Studies in Sanskrit, Jawaharlal Nehru University, New Delhi, the Coordinator of the conference and to Sanjay Saxena, Senior

Advisor to Chancellor, Dev Sanskriti University, Hardwar. I offer immense thanks to Sri Balakrishna D. Kalbagh, a renowned astrologer and Lecturer in Sanskrit, National PU College, Basavanagudi, Bangalore, Prof. H. Leelavathi, Principal, National Degree College (Autonomous); Prof. H.K. Ambika, Principal, National PU College, Bangalore, for their suggestions. My special thanks to Smt Veena H.S., Secretary, The Indological Research Foundation, Bangalore, for type-setting, S. Chidananda, B.E. CSE, Bangalore Institute of Technology, Bangalore, M.P. Pavan, B.E. CSE, JSS Academy of Technical Education, Bangalore, for preparing the tables.

Endnote

My father Sri Melukote P. Krishnamurthy (16-12-1912-3-11-2006) was a renowned astrologer by hobby and prepared the horoscopes of the members of our families, relatives and friends. Many of his predictions written well in advance with respect to the horoscopes of family members, relatives and friends came true. He even suggested inexpensive methods for performing rituals by beneficiaries to avert mishaps, overcome difficult times, etc. Around June 1980, he told me that Jupiter (*guru*) along with Saturn (*śani*) are in ascendancy in my horoscope and that I would accomplish everything academically and in adventure sports for the next sixteen years, which became true. He clearly stated, my periods of foreign travels, assignments, and awards, etc. years in advance. When my mother died on 12-6-2002, the next day evening, he said to the family members that he would live for four years and odd months which became a reality.

Professor Lalithamba Chandrashekhar, formerly a Principal of a Teachers Training College and a family friend of ours, is a reputed astrologer by hobby. Like my father, she predicted about the auspicious and inauspicious events of my life and of the inmates of our family well in advance which became absolutely true. Her predictions regarding the education, health and nature of my son were crystal clear. These two never practised astrology for profit and were great worshippers of personal deities and spiritual seekers (*sādhaka*s) by themselves.

References

Jātakābharaṇam, Puṇḍit Dhuṇḍhi Rāja

Bhṛgu Saṁhitā, Bhṛgu

Bṛhat Saṁhitā, Varāhamihira

Siddhānta Jyotiṣa

Mānasāra

Kālamṛtam

Muhūrtta Dīpikā

Web articles

Web Sources: http://www.bartleby.com/59/17/typebpersona.html
http://www.alleydog.com/glossary/definition.cfm?term=Type%20B%20Personality
http://www.uncc.edu/pagoolka/TypeA-B-intro.html
http://www.longevitywatch.com/personal.htm

Table of Horoscopes Analysed in the Paper

Gen-der	D.O.B	Saṁvatsara	Māsa/ Month	Pakṣa	Day
M	14/12/1912	Paridhavi	Mārgaśira	Śukla	Saturday
F	9/10/1925	Krodhana	Aśvayuja	Kr̥ṣṇa	Friday
M	16/07/1950	Vikr̥ti	Āṣāḍha	Śukla	Sunday
M	10/2/1957	Durmukhī	Māgha	Śukla	Sunday
F	24/03/1963	Śubhakr̥ta	Phālguna	Kr̥ṣṇa	Sunday
F	19/5/1947	Sarvajit	Vaiśākha	Kr̥ṣṇa	Monday
M	6/7/1941	Viśu	Āṣāḍha	Śukla	Sunday
F	25/8/1975	Rākṣasa	Śrāvaṇa	Kr̥ṣṇa	Saturday
F	16/7/1951	Khara	Bhādrapada	Kr̥ṣṇa	Saturday
M	11/2/1932	Prajotpatti	Māgha	Śukla	Thursday
F	16/12/1977	Paiṅgala	Mārgaśira	Śukla	Friday
F	13/9/1942	Citrabhānu	Bhādrapada	Śukla	Sunday
F	27/11/1962	Śubhakr̥ta	Kārtika	Kr̥ṣṇa	Tuesday
M	4/2/1959	Vilambi	Puṣya	Śukla	Wednesday
F	22/8/1964	Krodhana	Śrāvaṇa	Śukla	Saturday
M	9/6/1921	Durmati	Jyeṣṭha	Śukla	Thursday
M	5/3/1919	Kālayukti	Phālguna	Śukla	Wednesday
F	16/3/1953	Vijaya	Jyeṣṭha	Śukla	Monday
F	13/9/1976	Maṅgala	Bhādrapada	Kr̥ṣṇa	Monday
F	5/5/1927	Prabhava	Vaiśākha	Śukla	Sunday
M	23/3/1920	Siddhārthī	Phālguna	Śukla	Monday
F	5/3/1931	Pramoduta	Phālguna	Kr̥ṣṇa	Thursday
F	14/6/1955	Manmatha	Jyeṣṭha	Kr̥ṣṇa	Monday
M	19/4/1960	Śarvari	Caitra	Kr̥ṣṇa	Tuesday
F	24/8/1957	Hevilambi	Śrāvaṇa	Kr̥ṣṇa	Saturday
M	11/8/1945	Pārthiva	Śrāvaṇa	Śukla	Saturday

M	5/2/1927	Akṣaya	Māgha	Śukla	Saturday
M	25/10/1906	Parabhava	Aśvayuja	Śukla	Tuesday
M	2/3/1938	Īśvara	Māgha	Kṛṣṇa	Wednesday
F	6/11/1943	Svarbhānu	Kārtika	Śukla	Saturday
M	5/3/1923	Dundubhi	Māgha	Kṛṣṇa	Monday
F	3/6/1929	Śukla	Vaiśākha	Kṛṣṇa	Monday
F	8/8/1933	Śrīmukha	Śrāvaṇa	Kṛṣṇa	Tuesday
M	6/7/1964	Krodhī	Jyeṣṭha	Śukla	Monday
M	8/2/1960	Vikārī	Māgha	Śukla	Monday
F	10/12/1954	Jaya	Mārgaśira	Śukla	Monday
F	30/11/1967	Plavaṅga	Kātika	Kṛṣṇa	Thursday
F	14/11/1949	Virodhi	Kārtika	Kṛṣṇa	Thursday
F	24/10/1948	Bahudhānya	Kārtika	Śukla	Monday
M	3/11/1936	Dhātu	Aśvayuja	Kṛṣṇa	Tuesday
F	3/11/1948	Sarvadhārī	Kārtika	Śukla	Wednesday
M	28/5/1965	Viśvāvasu	Vaiśakha	Kṛṣṇa	Friday
F	15/8/1968	Kīlaka	Śrāvaṇa	Kṛṣṇa	Thursday
F	9/2/1984	Rudhirodgārī	Māgha	Śukla	Thursday
F	7/2/1962	Plavaṅga	Māgha	Śukla	Wednesday
M	12/10/1984	Raktākṣi	Aśvayuja	Kṛṣṇa	Friday
M	29/5/1990	Pramoda	Jyeṣṭha	Śukla	Tuesday
M	9/6/1983	Vibhava	Jyeṣṭha	Kṛṣṇa	Thursday
F	7/11/1992	Aṅgīrasa	Kārtika	Śukla	Saturday
M	11/5/1994	Bhavana	Vaiśākha	Śukla	Wednesday
M	24/5/1995	Yuva	Caitra	Kṛṣṇa	Monday
F	15/1/2000	Pramathi	Puṣya	Śukla	Saturday
F	3/5/1952	Vandana	Vaiśākha	Śukla	Saturday

15

The Applicative Wonders of Āyurveda
The Mother of All Life Sciences

Vaidyaratnam R. Raghavan

In this paper we present the infinite possibilities of the Āyurveda as the mother of all life sciences. We emphasise the applicative aspects of Āyurveda by illustrating a number of seemingly incurable diseases which Āyurveda has successfully handled. Āyurveda with its patient centric approach and Vedic knowledge as firm foundation has remedy for any malady.

Introduction

Āyurveda is a holistic science of life. Being an Upaveda of *Atharvaveda*, the physical and metaphysical axiomatic principles enshrined in the Vedas become commonly applicable to Āyurveda too. This compendium of divine wisdom has been expounded by our *ṛṣi*s, the true visionaries of Vedic wisdom. For centuries, Āyurveda has borne the torch of hope in the management of diseases in the unknown corridors of *pratyakṣam* (perceivable by sensory knowledge) and *apratyakṣam* (extra-sensory) by merely applying the simple truth accentuated in its texts.

Vedic Foundation of Āyurveda

The Vedas are the oldest compendium of objective, subjective, and intuitive experiential knowledge. Conjunctionally, Vedas are considered to be *apauruṣeya* (not revealed by man). Here the knowledge is infinity and it delivers meaning to the listeners or seekers. The Vedic concepts are universal in nature. It is not influenced by *deśa* and *kāla*. The geographical barriers are not applicable to it and so also time. So being an Upaveda, it is implied that the Vedic epistemology and cosmology are fully acceptable to Āyurveda also. All Vedic philosophical explanations in so-called *saḍdarśana*s — Sāṁkhya, Yoga, Nyāya, Vaiśeṣika, Pūrva Mīmāṁsā, Uttara Mīmāṁsā, are applicable to Āyurveda also. That simply means, Vedic knowledge systems are to be enjoined for the proper understanding and practice of this life science. Then only it will become Vaidika. The Vedic concept of the universe is totally applicable to central point of creation — the man. The concepts of *brahmāṇḍa* — the universe, the *pinḍāṇḍa* — the man, are related. The former is the macrocosm, the latter the microcosm. This unique interrelation of man and universe is the firm theoretical foundation of Āyurveda on which the entire edifice of this life science is built.

Profundities of Āyurveda

True Āyurveda envisages a patient-centred therapeutic approach, to restore perfect health by eliminating all the pathogenic factors embedded in physiological and psychological realms, safely. The great generalisation of health and disease are basis of promotive and curative methodologies of Āyurveda. The fundamental considerations of Āyurveda are:

1. *Hetu* — basic cause,
2. *Liṅga* — the symptomatology,
3. *Auṣadha* — remedial measures.

Tridoṣas — *vāta, pitta, kapha* are irreducible absolute of existence. The balance of the *tridoṣa*s is health. Any imbalance of *tridoṣa*s is disease. The duty of the Āyurvedic physician is to:

1. protect the health by promoting techniques
2. safely cure the diseases by re-establishing the balance of *tridoṣas*.

Āyurvedic pulse examination is the most precise tool of diagnosis. It can fully decipher the description of pathogenesis in *Nidhāna* (Pathology) to equip the physician to choose the appropriate therapeutic procedures for the patient and to even foretell the prognosis. Pulse — the orchestra of one's life — reveals the following details among many other things:

1. State of the embryo in mother's womb.
2. Constitutional bend of the patient as paternal, maternal, or combination of both.
3. Evolutionary stages of neo-natal, infancy, childhood.
4. Pleasant and unpleasant periods of life undergone by the patient.
5. Psychological and emotional make-up of the patient
6. Cognitive power.
7. Status of health from past to present day.
8. Examination during sleep on vital days will reveal *pūrvajanma* too.

In Āyurveda the central point of consideration is the patient whereas in modern system of medicine many other factors are also essential for the developing therapeutic measures. Already we are having more than seven thousand syndromes; of course many more are to be encountered in future. In many conditions, the modern system of medicine is groping in the dark. In the recent times, growing resistant of bacteria is another global threat. This contingency will never arise in Āyurveda because it is patient-centred science. In this concept, every aspect of health and disease is amalgamated. Hence any malady has got remedy in Āyurveda. All attempts of academic modern science, especially life sciences, are acceptable to Āyurveda too, to the extent of agreement of *śruti, yukti,* and *anubhava*. Like the Vedas, Āyurveda also has got very liberal and catholic approach to existing knowledge system. The Āyurvedic physician must be well informed of existing knowledge system and keep a liberty to apply that knowledge. Āyurveda is to be enriched with life support system of the modern science such as diagnosis laboratory investigation and all other advanced techniques of diagnosis, for example, CT scan, MRI scan, ultrasonography. By enjoining these techniques Āyurveda will provide successful emergency management procedures which are not yet thought of. Our heritage is successfully applying this for last four decades. The unique individuality of Āyurveda is that it remedies each individual differently through the complex symptoms and the cause of a disease might be shared commonly with many others. This sets it apart from all other systems of medicine including modern medicine which targets the disease and treats uniformly all persons having the same disease. This is the cause why modern remedial measures are discarded periodically, while encountering disastrous side-effects. This can never happen in patient-oriented therapy of Āyurveda. For antibiotics Āyurveda has better substitute. Āyurveda can contribute a lot in treatment of malignancy, preventive oncology, and palliative care.

Our Dattātreya Heritage

The growth of Āyurveda was adversely affected by many socio-political and cultural factors for the last several centuries in our country. Fortunately Dattātreya Heritage, the oldest Āyurvedic order, has uninterruptedly carried the dynamic process of this science through *svarūpa rakṣā* (preservation of the literature in its original form), *artha rakṣā* (preservation of the intended meaning of the literature) and *prayoga rakṣā* (the application of the principles enumerated in the Śāstras). Through our heritage's continuous response to the present day health challenges of mankind, we have been able to equip ourselves in handling disease prevention (*svastha vṛtta*), cure of rare diseases (*adhura vṛtta*) and expand it to the management of emergency conditions, industrial disasters, epidemics and traumatic neurological conditions.

Infinite Possibilities of Āyurveda

We are now at a juncture where as representatives of the most ancient knowledge system, we need to prove to the world the proclamations of our *ācārya*s (seers) that "Āyurveda has remedies for all maladies." This normative science carrying the ancient insights should be appropriately integrated with the advances in science and technology to successfully manage diseases and to deliver evidence-based results.

Here, we would like to present some representative cases which we have successfully handled which are hitherto unknown possibilities of Āyurveda, thus accentuating the applicative wonders of the mother of life-sciences:

1. *Multiple Sclerosis* — perceived diseases of the West and an unpredictable and common neurological disorder in young adults which is characterised by benign to disabling and devastating symptoms. This has no cure in modern medicine.
2. *Atypical Hemolytic-uremic syndrome, Hemolytic anemia and HUS* — which can be genetic, acquired or idiopathic and is characterised by fatal hematological and nephrological complications and is associated very poor prognosis in modern medicine.
3. *Epidemic management of Chikungunya and other viral diseases* — mostly appear in epidemic proportions. Many crippling complications occur as sequel attaining chronicity. In all these stages modern system of medicine can provide only symptomatic remedy with a high probability of several, systemic complications.

4. *Guillian Barre syndrome and an autoimmune disorder* — which is characterised as ascending paralysis developing from few hours to 3-4 weeks. It can be life-threatening. Modern medicine has no cure, but manages the syndrome with therapies.
5. *Hemo-cromotosis* — is an inherited liver disease of the West leading to abnormal accumulation of iron in parenchymal organs thus leading to organ toxicity, liver cirrhosis, heart failure, and diabetes mellitus and to death. This disease has no cure in modern medicine and the procedure is hazardous with blood-letting routinely and the sickening side-effects make life impossible.
6. *Sarcoidosis* — an abnormal immune response displaying symptoms of specific inflammation of the body tissues which can heal by itself or can fatally harm the function of vital organs like lungs, heart, nervous system, liver, heart or kidneys. Modern medicine treats it with corticosteroids with a high possibility of recurrence within three years and in advanced cases the reversal of fibrosis in the lungs is not possible.
7. *Chronic obstructive pulmonary diseases/industrial lung diseases/obstructive sleep apnea* — COPD is the fourth major cause of death in USA and is characterised by progressive difficulty in breathing as in both emphysema and chronic obstructive bronchitis. Modern medicine doesn't have a cure and is not aware how to reverse the lung damage.
8. Diabetes micro-vascular complications can be reversed with Āyurvedic treatment.
9. Aplastic anaemia.

There are 100s of such diseases which are rare, not treatable by modern medicine but which can be handled by

Āyurveda. We have limited ourselves here with some representative cases.

Conclusion

World Health Organisation treats Āyurveda as "An alternative system of medicine/complementary medicine". But, the richness, efficacy and supremacy of Āyurveda are at least equal or even better than modern medicine. This is due to holistic approach taken by the Āyurveda and the solid foundation provided by Vedic wisdom. Āyurveda should provide the suffering mankind the safest remedy for leading a·productive life and achieving the ultimate goal of life — *mokṣa* (liberation). It is the need of the hour that we realise the infinite possibilities of this healing science and we refuge ourselves under the sanctuary of this Vedic science to improve the overall state of world health.

References

Sharma, P.V., 2000, *Caraka Saṁhitā* , Varanasi: Chaukhambha Orientalia.

Bhishagratna, K.K., 1999, *Suśruta Saṁhitā*, Varanasi: Chaukhambha Orientalia.

Hari Shastri, P., 1984, *Aṣṭāṅga Hṛdayam*, Nagpur: Krishandas Academy.

Babior, B .M., 2000, "The Megaloblastic Anemias", in *Williams Hematology*, New York: McGraw.

Lux, S.E., Palek J., 1995, *Blood: Principles and Practices in Hematology*, Philadelphia: Lippincott.

Parry, G.J., J.S. Steinberg, 2007, *Guillain-Barre Syndrome: From Diagnosis to Recovery*, Demos Health.

Raine, C.S. et al., 2008, *Multiple Sclerosis: A Comprehensive Text*, Saunders.

Agrawal, M.M., 2001, *Six Systems of Indian Philosophy*, Varanasi: Chaukhambha Orientalia.

16

Āyurvedic Medicines
Ancient Roots and Scientific Validation

Yamini Bhushan Tripathi

Science and technology in Ancient Indian texts is very rich and needs to be re-explored by scientific tools, so that new generation may understand and appreciate the same. Since science is objective, so it is best approach to achieve this goal. Āyurvedic system of Indian Medicine is in practice since centuries and for revalidating the clinical claims about its basic concepts and drugs, good research is needed. We have explored the mechanism of action of *Arjuna* (*T. arjuna*), *Bhilawa* (*S. anacardium*), *Guggul* (*C. mukul*), *Puṣkaramūla* (*I. racemosa*), which are effective in heart disease. We have also found that all Āyurvedic drugs are not safe, and they are without any side-effect. Here, new methodology for purification of *bhilawa* nuts has been developed, as unpurified nuts are highly toxic. In Western countries these Āyurvedic drugs are marketed as food supplement, so we have developed polyherbal formulation for heart disease (BHUx), diabetic nephropathy (PTY-1) and radiation-induced cellular damage (Y-Rad-1). These products are safe, without any contamination with heavy metals, insecticides, pesticides residues and pathogenic microbes, which is of great concern to any consumer. In this connection, "BHUx" has been patented in USA (Patent No. 7,416,743 B2 dated 26, August 2008), European Union (Patent No. 1583499 dated 25 June 2009) and China (Application No. 2003380109770.X dated 20 July 2007). It is a medicine which can be used for prevention and progression of atherosclerosis. Similarly Y-RAD –A is used to prevent radiation hazards (ERIP/0201022/M01 dated 3 April 2003) and PTY-1 is a single pure compound, which has been developed to prevent diabetic nephropathy through inhibition of PKC (patented in India-2751/DEL/2008 dated 04 December 2008). I have also proposed a hypothesis about a correlation between *tridoṣa* theory of Āyurveda and cellular biochemical steps[1], which is required to decode the secret of Āyurveda.

Introduction

SCIENCE and Technology in Ancient Indian texts is very rich and needs to be re-explored by scientific tools, so that new generation may understand and appreciate this traditional knowledge. This is because, science is objective and it could be understood by all stakeholders. Demand of natural products is continuously increasing worldwide, because of its eco-friendly nature and lesser side-effects, than the popular system of conventional medicine. People are aware about the benefits of the natural products and therefore they are deeply interested in its use, especially, in the first stage of disease, before visiting a doctor. In other words, they wish to use it as preventive medicine on priority basis, but they are not fully confident about their mechanism of action and associated side-effects. In this situation, demand of scientifically validated food supplements/functional foods/nutraceuticals is high. Therefore a consolidated effort should be made to standardise these products, and to provide more information regarding their efficacy, toxicity and herb-drug interaction. The people are more concerned about toxicity on long use, adulterations, contamination with heavy metals, pathogenic micro-organisms and residual of pesticides and insecticides. These issues could be targeted easily by developing a strong working group in collaboration with leading scientists in these fields, for more research and by taking more precautions in manufacturing and packing of these products.

---

1. Y.B. Tripathi, 2000, "Molecular Approach to Ayurveda", *Indian J Exp. Biol*, 38:409-14, 2000.

Another important issue is the shelf value of these products. Since these are natural products, so it is difficult to keep them for a long time. It would be always preferred to prepare them freshly, or to use them within a short span of time. This information must be generated with each product and should be labelled properly on the products for their better acceptability and marketing.

Asian countries, including India and China, are the major players in marketing of these herbal products in the Western developed countries, but India has very low share of about 1.5 per cent, in the international herbal export as compared to its competitor, China, who has about 25 per cent share of this market. This is a point of great concern that in India, there are about 8000 pharmacies, which are involved in manufacturing herbal drugs and we have large biodiversity, but still our share is so low.

When we talk of standard products, then there are two ways to standardise them. One is the biological method, through which one can fix the therapeutic index of a product for its claimed properties as described in the ancient texts. This is useful for polyherbal formulations, a mixture of many plants, such as *cyavanprāśa* or *daśamūla kvātha*. The second option is the chemical standardisation, where one can do fingerprinting of medicinal plants, through thin layer chromatography, gas chromatography or column chromatography, named as TLC, HP-TLC, HPLC, GLS, GC-MS, LC-MS, etc. Besides, there is a need to evaluate the contamination levels, as described above, by using best instrumentation facilities.

Further, when we talk about mechanism of action then different *in vivo* and *in vitro* models should be used to understand its signalling cascade, target enzymes in cell and also the active molecule for biological activity. Proteomics, genomics, pharmaco-dynamics, tools of molecular biology, etc.

must be used to explore these aspects. Besides the animal experiments, cell culture, tissue culture and organ culture could be used to understand this aspect of drug development.[2]

At this stage, it is important to note that besides the facts described above, proper marketing and documentation of the available data is also essential to enhance Indian share in the export market of herbal drugs. At present, several countries are making new laws to regulate the marketing of nutraceuticals in their countries, such as FDA-USA guidelines. Till recent past, the marketing of herbal products was not well regulated and it was in the hands of natural healers, who were free to import these herbal preparations from other countries. However, several reports of toxicity and adverse effects were reported from time to time, by using these herbs, when taken alone or in combination of allopathic medicines. Although, these reported cases were not statistically significant, as compared to large number of users but still this made a great issue in its regulation.[3] The main issue, which is of great concern to all the countries, who are major players in this field, is related to the inclusion of medicinal plants in that list, which allows for export and import of these herbal products as food supplement, under WTO agreement. This issue is being raised by several activists and NGOs in international submits, to bring awareness about the benefit of these herbal formulations and for liberalisation of rules, but still success has not been achieved.

The pharmaceutical houses of modern drugs are also not in favour of free marketing of herbal products in name of nutraceuticals. They desire to have strict regulation as it is

2. Yamini B., Tripathi, and Deepshikha, Tripathi, "Need to Clinically Investigate BHUx for Atherosclerosis and Osteoporosis", *Current Science*, vol. 93, (12), 1658, 2007.

3. Yamini Bhushan Tripathi, "Challenges in Marketing Ayurvedic Drugs in USA : Possible Remedies", *Current Science*, vol. 95 no. 10, p. 1382, 2008.

existing for developing an allopathic drug, which takes a long time and huge money for conducting preclinical toxicity study and several steps of clinical trial, before a product comes to market as prescription drug.

There is another important aspect of Āyurvedic medicine, which is still untouched, is related to metallic preparations named as *bhasmas*. In ancient time, these metallic preparations were in medical use, but slowly as knowledge grew up about the toxicity of these heavy metals, the use of these products was restricted in various countries[4] of Saper et al.[5] However, again the interest in metallic preparations is developing with the emergence of nanomaterials from metals. It has been well documented that any material when converted to its nano size, changes its property.[6] It may be true to these *bhasmas* also. Therefore this knowledge of nanotechnology must be extrapolated to our ancient knowledge of *bhasma* preparation and their therapeutic uses.[7]

The Āyurvedic system of Indian Medicine is in practice since centuries and for revalidating its clinical claims, scientific research is required to understand the truth behind the basic concepts of Āyurveda, i.e. *tridoṣa* theory. Life (*āyu*) is the combination (*saṁyoga*) of body, senses, mind and reincarnating soul (*Caraka Saṁhitā*, Sūtrasthāna, 1.42-43.). Āyurveda is a

---

4. Robert, B. Saper, Stefanos N. Kales, Janet Paquin, Michael, J. Burns, David M. Eisenberg, Roger B. Davis, S. Phillips Russell, "Heavy Metal Content of Ayurvedic Herbal Medicine Products", *JAMA*. 292:2868-73, 2004.
5. Robert B. Saper, "Heavy Metals in Āyurvedic Medicines", *JAMA*, August 27, 2008.
6. Z. Nie, K.J. Liu, C.J. Zhong, L.F. Wang, Y. Yang, Q. Tian, Y. Liu, "Enhanced Radical Scavenging Activity by Antioxidant-Functionalized Gold Nanoparticles: A Novel Inspiration for Development of New Artificial Antioxidants", *Free Radic. Biol. Med*, 43(9), 1243-54, 2007.
7. Y.B. Tripathi, "A Multidisciplinary Approach to Standardize Bhasmas (Āyurvedic Metallic Preparation), *Current Science*, 90 (7), 2006.

holistic healing science, which comprises of two words, *Āyu* and *veda*. *Āyu* means life and *veda* means knowledge or science. So, the literal meaning of the word Āyurveda is the science of life. It is a science dealing not only with treatment of diseases, but deals with complete way of life. Its main object is to maintain the health of a healthy person, to prevent the onset of disease and also to treat a disease.

Āyurveda is thought to be originated from *Ṛgveda*, one of the oldest philosophical texts of the world (about 6,000 years). According to *Caraka Saṁhitā* (1.46-47), life is the outcome of equilibrium between body, mind and soul, which are like a tripod for a living entity. Although there is a long list of drugs, which could be tried for revalidation, but every laboratory and every scientist has a limitation for one's expertise and facilities. However in this process, we are describing here, those medicinal plants, which are directed towards age-related chronic disorders.

One can name these diseases as metabolic disorders or disorders related to food and lifestyle. In Āyurveda they are grouped under vātic disorders, which are now being compared with diseases due to initiation of inflammation or high free radicals. In fact, in Āyurveda, there are three *dhātus* (*vāta, pitta* and *kapha*) at somatic level and three *dhātus* at mental level (*sattva, rajas* and *tamas*). If these factors are disturbed or under imbalanced situation, then there is pathogenesis of a disease. Based on these *dhātus/doṣas*, all the diseases have been classified into three major groups. Combination of imbalance of more than one *doṣa* has been further sub-classified into more groups of diseases.

In fact, the verses of *Ṛgveda* consider five elements as the basic components for life. The same is true with Āyurveda also. These components include (1) space, (2) air, (3) earth, (4) fire, and (5) water. However, Sāṁkhya, an Indian philosophy, considers 24 basic elements responsible for existence of life. These five basic elements come together to create three different types of *doṣas*/constitutions, i.e. *vāta, pitta* and *kapha*.

If air and space is predominant then a person is of *vāta* constitution. Similarly, if water and earth are predominant, then one gets *kapha* constitution and if fire and water are predominant then one acquires *pitta* constitution. Now this concept of constitution is gaining support from human genome mapping, which encourages the concept of customised prescription, as it was already there in Āyurveda.

Going back to history of Āyurveda, this object of health science was delineated into two distinct schools: (1) Ātreya — The School of Physicians, and (2) Dhanvantari — The School of surgeons. These two schools wrote two major books on Āyurveda, i.e. *Caraka Saṁhitā* and *Suśruta Saṁhitā*. Later on around CE 500, Vāgbhaṭṭa compiled the third major treatise on Āyurveda, *Aṣṭāṅga Hṛdaya,* which contained knowledge comprising both the schools of Āyurveda. Later on sixteen major Nighaṇṭus or supplementary texts on Āyurveda were written, which compiled all existing information of respective periods.

After mapping of human genome, customised medication, based on the constitution of a person is gaining importance. In Āyurveda it is in practice since centuries. They use specific prescription for specific individual for same disease. In fact, the selection of drug for treating a particular pathology is dependent on constitution of that individual. It depends on constitution or personality of an individual. A man of *kaphaja* constitution will be prescribed with a list of medicine, which would be different than the prescription, for a person with same disease but with *vātaja* constitution.

Regarding prediction of constitution, there are several publications in the internet. It is through physical examination including pulse assessment and also by knowing the food habits and liking and disliking for food and climate. To summarise that, a person with thin build and small-boned with prominent joints, dry, chaps easily, rough, cool skin; prominent veins; coarse and dry hair, lean and thin body weight is considered as *vāta* constitution. Vātic imbalance may be regulated by eating warm foods, by avoiding extreme cold, raw and frozen foods.

Further, a person with *pitta* constitution would be of strong will-power, with good appetite and digestion power. Imbalanced *pitta* is manifested as skin rashes, burning, inflammation, fever, ulcers, anger, jealousy, copious urine and for its regulation one should prefer cool environment, avoiding excess heat, steam or humidity. They should also avoid excessive use of oily and fried foods as well as caffeine, alcohol, red meat, hot spices, or salt, choosing instead to eat fresh fruits and vegetables and whole grains.

Similarly, a person with *kapha* constitution is identified with good endurance, strength and stamina. These persons have a tendency towards congestion, sinusitis, sluggishness, weight gain, and diabetes or water retention. For its balancing one should enhance physical activity, consume little fried or fatty foods, and avoid cold drinks, sweets or excessive amounts of bread. They should enjoy plenty of fresh vegetables. I have proposed a hypothesis to explain the basic concept of *tridoṣa* theory of Āyurveda through cellular biology tools[8] which is required to decode the secret of Āyurveda.

The treatment part is also classified according to constitution, with a holistic approach. For example, there are about 80 diseases, which have been classified as *vātaja* disorders, under a separate chapter of "Vāta Cikitsā" in *Caraka Saṁhitā*.[9, 10] These diseases are mainly due to imbalance of *vāta*.

8. Tripathi, Y.B., "Molecular Approach to Ayurveda", *Indian J Exp. Biol*, 38:409-14, 2000.

9. Kashinath Pandey, Gorakhnath Chaturvedi, *Caraka Saṁhitā*, Part-II: Chapter 28, Vāta Vyādhi Cikitsa, p. 775, Chaukhambha Surabharati Academy, 1998.

10. Kashinath Pandey, Gorakhnath Chaturvedi, *Caraka Saṁhitā*, Part-I: Chapter 12, Vāta Kalakali, p. 243, Chaukhambha Bharati Academy, 1998.

Although in modern medicine, all these 80 diseases are treated separately for their pathogenesis and treatment, but in Āyurvedic texts, they are considered as one and thus the treatment is also focused to bringing back the imbalance of disorder of *vāta dhātu*.

In this series a long list of medicinal plants has been explored in our laboratory, with special reference to their action as hepato-protective, hypo-lipidemic, anti-atherosclerotic, anti-anginal, anti-inflammatory, anti-osteoporotic, anti-diabetic, anti-arthritis, radio-protective, immune-modulator, anti-cancer. Most of them have been investigated at the level of their target enzymes to understand their signalling pathway and their role on free radical generation system. Here, we have described some of them with their roots in ancient texts of Āyurveda.

Arjuna (T. arjuna)

The bark of *Terminalia arjuna* is in clinical use for the management of cardiovascular disorders in various systems of Indian medicine as Āyurveda, Siddha and Homoeopathy. In *Caraka Saṁhitā* decoction of bark of mature tree of *Terminalia arjuna* in milk, has been recommended for management of weak heart. It is a drug of choice by Āyurvedic physicians for various ailments of heart as described in *Bhāvaprakāśa Nighaṇṭu*.[11] However recent scientific studies have mentioned its role as hypolipidemic,[12] hypoglycemic.[13] We have reported it as the Ca Channel blocker, as its mechanism of action for its

11. G.S. Pandey, K.C. Chunekar, *Bhāvaprakāśa Nighaṇṭu*, vol. 1, Varanasi: ChaukambhaVidya Bhavan, 1998.
12. H.P. Shaila, S.L. Udupa, A.L. Udupa, "Hypolipidemic Activity of Three Indigenous Drugs in Experimentally Induced Atherosclerosis", *Int J Cardiol.*, 67(2):119-24, 1998.
13. S. Dwivedi, A. Aggarwal, "Indigenous Drugs in Ischemic Heart Disease in Patients with Diabetes", *J Altern Complement Med.*, 15(11):1215-21, 2009.

role in heart disease and for management of hypertension. In a study, we studied the effect of various organic fractions of *Terminalia arjuna* on potassium chloride (160 mM) induced contraction in smooth muscle and our results indicated that the petroleum ether extract did not exhibit as good response in case of its alcoholic extract.[14] We further reported that its alcoholic extract modulates the contraction of rat aorta induced by norepinephrine also. The extract of *Terminalia arjuna* relaxed the contraction, induced by both KCl (60 mM) and norepinephrine (NE, 3 μM) in the concentration dependent manner. The response was comparable with diltiazem, a standard calcium channel blocker (Dz, 1.0 μM), which showed inhibition up to 68 per cent of the KCl induced and 30 per cent of NE induced contraction. In similar conditions *Terminalia arjuna* extract showed inhibition up to 80 per cent in the NE induced contraction and 30 per cent in KCl induced contraction.[15]

Bhilawa (S. anacardium)

Semecarpus anacardium has been in use for a long period in Āyurveda and Siddha. The nuts of *bhilawa* tree is used as medicine. Previously, the sap of the nut was dried as black resin, which was mixed with lime or alum and used as a dye or marking-ink; hence the name came as marking nuts.

Its tree is found throughout the hotter parts of India and in the deciduous forests of the Malaysian archipelago and northern Australia. Its tree is of medium-size, reaching up to a height of 12-15 m, with dark brown and rough bark. The leaves are large, ovate and simple up to 60 cm long and 30 cm wide. The flowers are small, dioecious, dull greenish-yellow;

14. Y.B. Tripathi, V.P. Tripathi and P. Tripathi, "Effect of T. arjuna-extract on KCl Induced Contraction on rat was Deferens", *Phytotherapy Research*, 3 (4) 162-164, 1989.

15. Y.B. Tripathi, S. Pandey, P. Tripathi and M. Sharma, "Antioxidant Property of R. cordifolia : Comparison with Vitamin E and Parabezoquinone", *Phytotherapy Research* 9 (6) : 440-43, 1995.

the fruits are drupes about 2.5 cm long, obliquely ovoid, smooth and shining, black when ripe, on a fleshy orange receptacle in terminal panicles. Its fruit, gum and oil have been in use for long time in India and other Asian countries. The resinous juice is applied to heel cracks. In traditional medicine, it is highly valued for the treatment of tumours and malignant growths. The fruit is reported to be caustic, astringent, alterative, antirheumatic, carminative, counterirritant, rubefacient and vesicant.

Traditionally *S. anacardium* is used for rejuvenation, rheumatoid arthritis, fever and neurological disorders. It is also used in anorexia, cough, asthma, indigestion, enlargement of the spleen, alopecia, ulcers, corns, leprosy, leucoderma, rheumatism, piles and for various nervous diseases. In animal care also it has a therapeutic use. Its crushed seeds are applied to cuts, bruises and scratches to stop bleeding. The oil is applied to cracks and fissures in the tails and hooves of animals. *Bhilawa* fruits are cooked in mustard oil and then applied to sores to kill maggots. From chemistry point of view, its seed contains anacardoside and other derivatives of anacardic acids, including *bhilawanol*, which is a mixture of 1, 2 dihydroxy-3 (pentadecenyl-S′) benzene and 1, 2-dihydroxy-3-(pentadecadienyl-S′, 11 ′) — benzene. A few flavones, namely semecarpetin, semecarpuflavone, nallaflavone, galluflavanone, jeediflavanone, tetrahydroarmentoflavone and tetrahydrobustaflavone have been isolated from its fruit. It is also a rich source of fixed oils, which mainly contains palmitic, stearic, oleic, linoleic and arachidic acids. From Āyurvedic text it has following medicinal classifications based on their *guṇa*s

Guṇa: *laghu* (light), *snigdha* (unctuous), *tīkṣṇa* (sharp)

Rasa: *madhura* (sweet), *kaṣāya* (astringent)

Vīrya: *uṣṇa* (hot)

Vipāka: *madhura* (sweet)

Doṣa: pacifies *vāta*

Various experimental and clinical studies have supported its several ancient claims, which are available in scientific literature with full references. Some of its pharmacological properties include its antineoplastic activity, radioprotective, immunomodulatory activity, antiarthritic activity, antilipid peroxidative, anti-inflammatory, antioxidant, anthelmintic activity and hypocholesterolaemic activity.

However, Āyurvedic texts advocate its use only after proper purification, because of its associated toxic effects also, especially the anacardic acids, which are known allergic. Several traditional methods for purification are available, but they are not free from toxicity and also the scientific basis behind these purification steps is not known. Recently, we developed a new method for purification of SA nuts. Instead of using brick powders (traditional method of Āyurveda), we used silica gel. Parallel to this technique, we also removed the hexane soluble fraction of the nut as the part of purification. Based on our studies in macrophage culture, we could suggest that oily part of the nuts is pro-oxidant, whereas the alcoholic fraction is anti-oxidant, and that the use of hexane is better for purification of SA nuts than the traditional methods, as it enhances the therapeutic value. We used a biochemical parameter as its effect on nitric oxide production or prevention of lipopolysaccharides (LPS) induced NO production, for testing the biological activity of these isolated fractions.[16] Our another report showed that a fraction of *S. anacardium* at dose of 1 mg/100 g body wt, significantly reduced serum cholesterol from 378.87 mg/dl in the rats fed with atherogenic diet (AD) to 197.99 mg/dl (45-52 per cent) in the rats fed with AD diet

16. Yamini B. Tripathi, Nidhi Pandey and Pratibha Tripathi, "Purification of Nuts of Semecarpus Anacardium: A Herbal Drug for Arthritis", *Current Science*, vol 94, no. 8, 2008.

and increased serum HDL-cholesterol (33-37 per cent). The same fraction also inhibited LPS induced NO production in the culture of activated rat peritoneal macrophages in the dose dependent manner with IC50 value at 50 ng/ml of the culture medium. The drug in the above doses was completely safe and non-toxic (no change in the enzymes), to liver and kidney functions.[17]

We have also tested the effect of alcoholic extract of SA nuts, which has been purified by traditional methods on cell cycle of DU-145 cells. It significantly induced the cell cycle arrest at the level of cytokinesis, suggesting its therapeutic role in management of prostate cancer.[18] In the process of developing new drugs from SA combinations, we tested the effect of *sandhika*, a polyherbal formulation (water soluble fractions of *Commiphora mukul, Boswellia serrata, Semecarpus anacardium* and *Strychnos nux vomica*). It has been in clinical use in India for last 20 years. Its modified formulation BHUx has shown specific inhibition of cyclooxygenase (COX)-2 and lipoxygenase (LOX)-15 and has been claimed to prevent the diet-induced atherosclerosis in rabbits.

It was further reported that *sandhika* significantly inhibited the adverse effect of lipopolysaccharide (1 microg/ml, LPS) on MC3T3-E1 osteoblast-like cells with reference to calcium nodule formation and alkaline phosphatase activity. After 10 days of treatment, it was found that *sandhika* maintained the calcification in these cells in concentration dependent manner when tested in range of 0.5 to 16 mg/ml final concentration in culture media. The cells were stained with "Alizarin S" for visualisation of calcium nodules and Meta Morph software

---

17. Y.B. Tripathi and R.S. Pandey, "Semecarpus anacardium L. Nuts Inhibit LPS Induced NO Production in Rats Macrophages along with its Hypolipidemic Property", *Indian J. of Expt. Biol.* (CSIR) 42:342-46, 2004.
18. Y.B. Tripathi, P.T. Reddy, M.V.R , S. D.S.T. Dutt, & E.P.K. Reddy, "Effect of Semecarpus anacardium on Cell Cycle of DU-145 Cells", *Phytomedicine* 5(5), 383-88, 1998.

(Universal Imaging, Downingtown, PA) was used for image analysis. This indicated the use of *sandhika* for management of osteoporosis, which is enhancing at an alarming rate in all developed and developing countries and has immerged as one of the major health problems with associated financial burden on the society.[19]

We also explored its radio-protective and chemo-therapeutic potential of a herbal formulation, which was patented it as Y Rad A. It is a combination of a selected fraction, of *Rubia cardifolia*. Y-RAD-A is used to prevent radiation hazards (ERIP/0201022/M01 dated 3 April 2003). It was tested in mice, exposed to Cobalt-60 radiation and response was reported in terms of bone marrow mitotic count, endogenous haematopoietic spleen colony formation (CFUs), micronuclei formation and blood leukocyte count.

Its effect was further tested on cancer management by using DMBA induced mammary carcinoma in rats and also in a cell culture system as pre-cancerous mammary alveolar lesions by incubating the mammary glands in glucocorticoids (1 mg progresterone) and estradiol in time span of 9 days, followed by supplementation with 5 mg insulin, 5 mg prolactin, 1 mg aldosterone and 1 mg hydrocortisone per ml of medium for another 10 days, along with different concentrations of these extracts. In this part of incubation 2 mg/ml DMBA was also added on third day and further incubated for 24 hours to induce the development of precancerous lesions. At the end of 10 days, the glands were transferred to a medium containing only insulin (5 mg/ml) for an additional 14 days. Finally the glands for alveolar lesion were fixed and stained with alum carmine for evaluation of un-regressed areas and evaluated for the incidence (number of glands with lesions/ total number of glands (*Indian J. Radiation Research*, 1 (2): 13, 2004).

19. Y.B. Tripathi, P. Tripathi, K. Korlagunta, S.C. Chai, B.H. Smith, B.J. Arjmandi, "Role of Sandhika: A Polyherbal Formulation on MC3T3-E1 Osteoblast-like Cells", *Inflammation*, 31 (1) : 1-8, 2008.

Guggul (*Commiphora Mukul*)

Guggul is a tree of small size and its resin is used as medicine. In Āyurveda it is mentioned as *vāta śāmaka* and pain reliever. It is also recommended as anti-obesity drug and for cleaning the arteries. "As per *Caraka Saṁhitā*, (Sūtrasthāna, Ch. 25), *guggul* is one of the best drugs among herbs that are used for obesity and *vāta* disorders." Similarly, Suśruta describes the use of *Commiphora mukula* for heart health. It is also used for fragrance, lightness, penetrating power, sharp, hot, mobile are the qualities in this gum. It is pungent in taste, bitter in post-digestion-taste.

Newly collected resin is unctuous, slimy, enhancing body weight and aphrodisiac, whereas old gum resin reduces the tissue mass and fat. It pacifies *kapha* and *vāta* due to sharp and hot properties. It expels out bile and excretory waste materials. As per *Bhāva Prakāśa* it has property of clearing, mobile, dry, light, penetrative, sliminess, with astringent, pungent in taste. It is bitter in post-digestion taste, hot in active principle and increases *pitta* and pacifies *kapha* and *vāta*. It promotes union of bone fractures, (*lākṣādi gug.*) and acts as tonic, aphrodisiac and promotes strength of body. It is useful in obesity, sore throat, wounds, urinary disorders, renal stones, skin disorders, acne, nodular swellings, edema, piles, lymphadenopathy and intestinal worms are helped with this gum.

Experimental details support its use for atherosclerosis (the ethyl acetate fraction), for increasing the plasma fibrinolytic activity, for lowering the serum cholesterol, triglycerides, as anti-arthritic and anti-inflammatory agent. It is also useful for activating the tyhroid gland specially useful in sub-clinical hypothyroidism. It has also shown enhancement of HDL cholesterol along with decrease in VLDL and LDL cholesterol.

It has shown antiseptic property with a bitter, stomachic and carminative property and for increase the appetite and enhancing the digestion power. It causes elevation of leucocytes in the blood and stimulates phagocytosis. It acts as a diaphoretic, expectorant and diuretic, and is said to be a uterine stimulant and emmenagogue. The resin is also used in the form of a lotion for corneal ulcers and as a gargle in chronic tonsilitis, pharyngitis, laryngitis and ulcerated throat.

Herbal extracts from *Commiphora mukul* (*guggul*) have been widely used in Asia as cholesterol-lowering agents, and their popularity is increasing in the United States. Recently, guggulsterones, the purified bioactive compounds of *guggul*, have been shown to be potent antagonists of two nuclear hormone receptors involved in cholesterol metabolism, establishing a plausible mechanism of action for the hypolipidemic effects of these extracts.

Effect of *Commiphora mukul* (gum *guggul*) in patients of hyperlipidemia with special reference to HDL-cholesterol has been reported. Some reports show the indication stomach discomfort or allergic rash, if taken in high doses. It should be avoided in pregnant or breast-feeding women and in children. Safety of use beyond four months has not been well studied. We have reported its antiatherogenic property through clinical trial in the name of *puṣkara-brāhmī guggul*,[20] BHUx.[21, 22] Its mechanism of action has been reported through stimulation of thyroid gland.[23]

---

20. B.N. Upadhyay, S.D. Sharam, V.K. Gupta, Y.B. Tripathi, "Role of Pushkara Guggulu in the Management of Ischaemic Heart Disease", *Ancient Science of life*, IV (1), 9-19, 1984.
21. Yamini B. Tripathi, Birendra K. Singh, Ravi S. Pandey, Mohan Kumar, *BHUx:* "A Patent Polyherbal Formulation to Prevent Atherosclerosis", eCAM, USA (Oxford Press), 2 (2) 217-221, 2005.
22. Y.B. Tripathi, BHUx: A Patented Polyherbal Formulation to Prevent Hyperlipidemia and Atherosclerosis, *Recent Pat Inflamm Allergy Drug Discov.*, 3(1):49-57, 2009.
23. Y.B. Tripathi, P. Tripathi, O. P. Malhotra & S.N. Tripathi, "Thyroid stimulatory Action of (Z) Guggulsterone: Mechanism of Action", Planta Medica 4:271-77, 1988. Germany.

Puṣkaramūla (*Inula Racemosa*)

Puṣkaramūla, also known as *Inula*, is used for: angina pectoris, heart disease, ischemic heart disease. It promotes normal diastolic blood pressure, liver conditions, cholesterol, respiratory tract conditions, asthma, bronchial disease, chest (precordial) pain, chronic bronchitis with cardiac complications, cough, dyspnoea, protects against bronchospasm induced by histamine, serotonin and pollens.

Its Sanskrit names includes *padma-patra* (as its leaves are like lotus petals), *kāśmīra* (as it generally grows in Kashmir area) and *kuṣṭhabheda* (as its characteristics and actions are like *kuṣṭha*) (*Saussurea lappa*). Its Lt. name is *Inula racemosa* of composite family. From Āyurvedic point of view its properties are as given below.

Rasa (taste) — *tikta* (bitter); *kaṭu* (pungent)

Guṇa (characteristics) — *laghu* (light)

Vīrya (potency) — *uṣṇa* (warm)

Vipāka (Post-digestion effect) — *kaṭu* (pungent)

In traditional text its uses have been described as *kāsa-śvaṣārapuṣkarmūla* is useful in cough and respiratory discomfort; *hikkā nigrahaṇa-puṣkaramūla* alleviates hiccough; *pārśvaśūlahara-puṣkaramūla* helps in pain in thorax region; *śophaghna-puṣkaramūla* is useful in all aedematous conditions; *pāṇḍunāśanam-puṣkaramūla* is useful in anaemia and its complications; *ardit vināśanam-puṣkaramūla* is useful in conditions involving nervous system specially the facial paralysis, *hr̥c-chūlaghna-puṣkaramūla* alleviates pain in heart region.

We have found that *Inula racemosa* root powder is very useful in patients with proven ischaemic heart disease and acts as adrenergic beta-blocking agent. The powder prevented ST-segment depression and T-wave inversion as observed in the post-exercise electrocardiogram. The petroleum ether extract of roots lowered plasma insulin and glucose levels within 75 min of oral administration to albino rats and it significantly counteracted adrenaline-induced hyperglycaemia in rats. The extract further showed negative inotropic and negative chronotropic effects on frog heart.[24]

Based on extensive research, we have developed some novel Āyurvedic drugs, which can be used as the food supplements for various diseases as described below. One of them is BHUx, which has shown to inhibit diet-induced atherosclerosis in albino rabbits, when orally given for three months. It showed remarkable reduction in intimal thickening of dorsal aorta in the treated animals. In addition, there was less calcification at the intima-medial interface and increased intensity of collagen cap on the surface along with an increase in survival, compared with the sham control.[25] It has significant anti-inflammatory properties through inhibition of cyclooxygenase-2 and lipoxygenase-15. BHUx significantly reduced inflammation in the carrageenan-induced rat paw oedema model of inflammation, suggesting its anti-inflammatory properties. It also inhibited cumene-hydroperoxide-induced lipid peroxidation (CHP) in liver homogenate, LPS-induced NO production in peritoneal macrophages and on key enzymes of arachidonic acid cascade, involved in the mediation of inflammation.[26]

---

24. P. Chaturvedi, S. Shukla, P. Tripathi, S.K. Singh & Y.B. Tripathi, "Comparative Study of Inula racemosa and Saussurea lappa on the Glucose Level in Albino Rats", *Ancient Science of Life*, 15(I): 62-70, 1995.
25. Nishant P. Visavadiya, and A. V. R. L. Narasimhacharya, *Asparagus* "Root Regulates Cholesterol Metabolism and Improves Antioxidant Status in Hypercholesteremic Rats, Evid Based Complement" *Alternat Med.*, 6(2): 219–26, 2009.
26. Y.B. Tripathi, M.M. Reddy, R.S. Pandey, J. Subhasani, O.P. Tiwari, B.K. Singh, P. Reddanna, "Anti-inflammatory properties of BHUx, A Polyherbal Formulation to Prevent Atherosclerosis", *Inflammopharmacology*, (UK) 12 (2), 131-152, 2004.

Finally, we have also developed a food supplement (PTY-1:) for management of diabetic nephropathy (patented in India-2751/DEL/2008 dated 04 December 2008). PTY-1 is a single pure compound, which has been developed to prevent diabetic nephropathy through inhibition of PKC and through anti-oxidant potential. It has been prepared from tubers of *Pueraria tuberosa Linn*. (PT), Leguminosae, which is a perennial climber, growing throughout tropical parts of India. In the Āyurvedic system of medicine, it is used as a drug of choice to manage pain, inflammation and other related diseases.[27]

Thus, it appears that Āyurveda is a knowledge of health care, which has developed on best scientific knowledge of that period including vast clinical survey and personal experience form *vaidya*s and users, but now it needs to be revalidated on modern scientific tools, because it is the language, which can be understood by all kind of people.

27. Nidhi Pandey, J. K. Chaurasia, O.P. Tiwari, and Yamini B. Tripathi, Antioxidant Properties of Different Fractions of Tubers from Pueraria Tuberosa Linn", *Food Chemistry*, 105, 219-22, 2007.

Part - VI

Text and Information Encoding in Human Brain and Computer

17

Unicode Standard for Vaidika Sanskrit

Efforts and Status

Swaran Lata
Manoj Jain

Abstract

The Vedas, written thousands years ago are vast source of knowledge. Vedas were mainly written in Vedic Sanskrit using Devanagari script. Veda texts were written using special symbols and characters to denote pronunciation/vocal nuances of a word. These symbols are placed around syllables, so that a reader can correctly pronounce/chant the verse. Unicode is de-facto industry standard for representation of multilingual text in electronic/IT and print media. To take the advantages of our ancient knowledge of Veda, it is very important that we should be able to bring these Vedas on ICT media. Department of Information Technology, Government of India, took initiatives to encode the Vedic Sanskrit symbols and characters in the Unicode Standard so that Vedas could be represented in computers/IT media. This report presents a brief overview of the efforts and current status of Unicode for Vedic Sanskrit symbols and characters. The code charts present various Vedic Sanskrit symbols and characters used in writing Vedas.

THE ancient holy text in India is known by the name of the Veda. The word *veda* is derived from the root verb *vid* — to know, to attain the knowledge. There are four Vedas, namely *Ṛgveda, Yajurveda, Sāmaveda* and *Atharvaveda*. Composed in verse format using Sanskrit language, the contents of these Vedas are of spiritual, philosophical and practical nature and are to be vocalised in terms of recitation, chanting, singing, etc. The collections of the *mantras* (the compositions in poetry and prose form) of the four Vedas were further divided into *śākhās* (branches). The range of Vaidika texts extends to their respective *śākhās*, *upaśākhās* (branches, sub-branches), Saṁhitās, Brāhmaṇas, Āraṇyakas, Upaniṣads, Śrautasūtras, Prātiśākhyas and Śikṣās. Originally, the Vaidika text (mostly in verse form) — Saṁhitās along with their respective *padapāṭhas* exist in the oral mode and were passed on from *guru* to *śiṣya* through various processes such as *śravaṇa* (listening), *anuccāraṇa* (repetition), *arthabodha* (understanding of the meaning), *kaṇṭhasthīkaraṇa* (memorisation), *carcā-saṁvāda* (discussion) and *praśnottara* (questions and answers).

The classes of reciters known as Vaidikas belonging to individual branches have retained their chanting characteristics while vocalising the *mantras*. At later stage such oral recitations had to be written down in order to retain the knowledge in the written mode. To write such Vaidika text with vocal nuances, various signs (*svarāṅkana-paddhati*) were devised and placed around syllables. These are called Vaidika accent and tonal signs.

There is special Vedic grammar, rules for each *śākhā* known as *prātiśākhya* and phonetic rules known as Śikṣā. There are four *upāṅgas*: Mīmāṁsa-Sūtras (described rules for interpretation of Vedic text), Nyāya and Vaiśeṣika Sūtras (deal with logical aspects, ontological classification, process of human understanding), Purāṇas (narrations of messages and teachings of Vedas), Dharmaśāstras (describe code of conduct for universal harmony). The Śikṣās cover the study of the terms such as *varṇa* (letters of alphabet), *svara* (accents like *udātta, anudātta* and *svarita*), *mātrā* (the time measurement required for the pronunciation), *bala* (stress — in context of *sthāna* and *prayatna*), etc. The *prātiśākhyas* constitute studies on phonetics, which prevailed in ancient India in different schools of the Vedas.

It is observed that the Vaidika text and its accent marks vary in their application and presentation as per its branch, sub-branch, tradition and location. It is interesting to note that a single chant, if vocalised by the Vaidikas of different branches, would have slight variation in terms of nasalisation, intonation, continuum, stress, etc. And to indicate these differences, many variants of a single sign have been developed. For example, there are 15 variations of *anusvāra* indicated by 15 graphic signs with slight variations in their visual forms. The selective usage of these signs indicates that the particular text is from a particular branch of Veda.

For example: In *R̥gveda*, the *udātta svara* is left unmarked. In *R̥gveda* and *Yajurveda* the *svarita svara* is marked with short vertical line on the top of a syllable. The same sign appears as *anudātta* mark below a syllable in *Atharvaveda* (Paippalāda Śākhā).

Sanskrit

Sanskrit is one of the most ancient languages of the world, which has molded the culture and the thought systems not only of India but also of many other countries in Asia. Sanskrit language is variously referred to as Devavāṇī, Amarvāṇī, Girvaṇavāṇī, Surbhāratī, Amarbhāratī, etc. each expressing connoting its inherent vitality, versatility and greatness. The script in which Sanskrit texts are written is called Devanāgarī. Sanskrit has its own place recognised by the linguists all over the world. It is agreed that the study of Sanskrit enables us to draw freely upon our tradition, which can lead to the new world outlook of modern man. Further, it helps to keep pace with the rapid social change, advances in modern science and technology and the process of modernization, at the same time inculcating the right type of social, moral and spiritual value through self-discipline.

Sanskrit is a living spoken language with a considerable literature of its own. Sanskrit is still spoken in some Indian families. Its vocabulary has permeated all Indian languages, and thus provides continuity with the past of our country. Pāṇini's book on Sanskrit grammar, named *Aṣṭādhyāyī*, has been considered by eminent American linguist Bloomfield as "one of the greatest monuments of human intelligence". Pāṇini was preceded by a long chain of grammarians, and his tradition continued even afterwards. With his 4000 *sūtra*s, each of which is usually no more than two or three words, Pāṇini was able to explain how almost all the words used in Sanskrit of his time were formed. It is precursor of today's generative grammar. Sanskrit grammar is prescriptive, that takes phoneme as smallest unit with meaning, knowledge representation is deeper and holistic at sentence level with three necessary and sufficient conditions of expectancy, compatibility and proximity whereas modern linguistics is descriptive and empirical, that takes morpheme as the smallest unit with meaning, uses word-by-word approach rather than sentence respectively. Sanskrit is syntax-free and word-order-free natural language. Śāstric Sanskrit is the natural language with all the desirable properties of formal artificial language, such as naturalness, expressiveness, unambiguous and no redundancy.

Unicode is the *de facto* industry standard for representation of multilingual text in computer/electronic media. Major scripts of the world are included in the standard including major Indian scripts. Unicode Standard is maintained by the Unicode Consortium and is in sync with ISO10646.

Deptt. of Information and Technology, Goverment of India (TDIL) is the member of Unicode Consortium with the objective of adequate representation of constitutionally recognised 22 languages (using 11 different scripts) in the Unicode Standard. TDIL programme is also co-ordinating the activities for representation/encoding of the Vaidika symbols and characters in the Unicode Standard. The Devanāgarī script is used for writing classical Sanskrit as well as Vedic Sanskrit. This includes the multi-tier usage of diacritic marks of complex compositions, above, below and at the sides of the base glyphs and various other symbols known as svarādi *cihna, visarga, bala, gomukha*, etc.

The task of identifying all such symbols, accent markers and characters used in Vaidika literature was very exhaustive. Prof. R.K. Joshi, CDAC Mumabi was entrusted the task of identifying all additional symbols/characters which were required to be encoded in the Unicode Standard. Prof. Joshi consulted many Vedic experts, studied many manuscripts, visited many institutes/libraries across India to find out various symbols which were needed for representation of Vedic Sanskrit in Unicode Standard. With his hard work and efforts the first draft was prepared in the year 2002. This draft was based on pure consonant approach and suggested encoding of the Devanāgarī pure consonants also, along with symbols/characters needed for Vaidika. The proposal was submitted to Unicode Technical Committee (UTC) in the year 2003. Suggested modification in the proposal as the proposed encoding of Devanāgarī pure consonants was not as per the Unicode Encoding Policies. Also, UTC advised to submit detailed evidences for the usage of these symbols/characters. Prof. Joshi and TDIL Programme worked to prepare the detailed evidences which were subsequently submitted to UTC.

The matter of encoding was further pursued with UTC in

May 2005. Mrs. Swaran Lata, Director, DIT, participated in the UTC meeting and presented/discussed the modified proposal on the encoding of the Vedic symbols and characters in the Unicode Standard.

Brown University, USA, also submitted another proposal to Unicode Consortium for the encoding of the Vaidika characters. The draft proposal submitted by Government of India and the proposal submitted by Brown University were again discussed during UTC#113. UTC recommended that Government of India will submit a unified revised proposal for encoding of Vaidika Sanskrit.

On behalf of DIT, GOI, Prof. R.K. Joshi, CDAC, Mumbai, further worked and prepared revised proposal (as advised by UTC) in consultation with the other experts in the area. DIT organised a number of meetings of Vaidika scholars and experts from all over India for consultations. The proposal was also uploaded on the TDIL programme website (http://www.tdil.mit.gov.in) for feedback and comments.

DIT submitted the final proposal to Unicode Consortium in January 2008 (L2/08-042 & L2/08-043) for consideration by UTC. The proposal was first discussed by the South Asia Sub-Committee and later by the UTC. The UTC recommended most of the proposed characters for inclusion in the future version of the standard.

UTC recommended encoding of various characters and symbols required for representation of Vaidika text in three different blocks, i.e. Devanāgarī, Vedic extensions and Devanāgarī extensions.

In the Vedic Extensions block (1CD0 – 1CFF), 35 characters have been recommended for encoding. 26 characters have been recommended for encoding in the Devanāgarī extended block (A8E0 – A8FF). Four characters were recommended for encoding in the Devanāgarī block (U+A8E0 – U+A8F9).

Devanāgarī (U+900 – U+97F)

	090	091	092	093	094	095	096	097
0	0900	ऐ 0910	ठ 0920	र 0930	◌ी 0940	ॐ 0950	ॠ 0960	॰ 0970
1	◌ँ 0901	ऑ 0912	ड 0921	ऱ 0931	◌ु 0941	◌॑ 0951	ॡ 0961	ॱ 0971
2	◌ं 0902	ऒ 0912	ढ 0922	ल 0932	◌ू 0942	◌॒ 0952	◌ॢ 0962	ॲ 0972
3	◌ः 0903	ओ 0913	ण 0923	ळ 0933	◌ृ 0943	◌॓ 0953	◌ॣ 0963	
4	ऄ 0904	औ 0914	त 0924	ऴ 0934	◌ॄ 0944	◌॔ 0954	। 0964	
5	अ 0905	क 0915	थ 0925	व 0935	◌ॅ 0945	◌ॕ 0955	॥ 0965	
6	आ 0906	ख 0916	द 0926	श 0936	◌ॆ 0946		० 0966	
7	इ 0907	ग 0917	ध 0927	ष 0937	◌े 0947		१ 0967	
8	ई 0908	घ 0918	न 0928	स 0938	◌ै 0948	क़ 0958	२ 0968	
9	उ 0909	ङ 0919	ऩ 0929	ह 0939	◌ॉ 0949	ख़ 0959	३ 0969	ॹ 0979
A	ऊ 090A	च 091A	प 092A		◌ॊ 094A	ग़ 095A	४ 096A	ॺ 097A
B	ऋ 090B	छ 091B	फ 092B		◌ो 094B	ज़ 095B	५ 096B	ॻ 097B
C	ऌ 090C	ज 091C	ब 092C	◌़ 093C	◌ौ 094C	ड़ 095C	६ 096C	ॼ 097C
D	ऍ 090D	झ 091D	भ 092E	ऽ 093D	◌् 094D	ढ़ 095D	७ 096D	ॽ 097D
E	ऎ 090E	ञ 091E	म 092F	◌ा 093E	◌ॎ 094E	फ़ 095E	८ 096E	ॾ 097E
F	ए 090F	ट 091F	य 0920	◌ि 093F		य़ 095F	९ 096F	ॿ 097F

Vedic Extensions (U+1CD0 – U+1CFX)

	1CD	1CE	1CF
0	◌᳐ 1CD0	◌᳠ 1CE0	ᳰ 1CF0
1	◌᳑ 1CD1	᳡ 1CE1	ᳱ 1CF1
2	◌᳒ 1CD2	◌᳢ 1CE2	ᳲ 1CF2
3	᳓ 1CD3	◌᳣ 1CE3	
4	◌᳔ 1CD4	◌᳤ 1CE4	
5	◌᳕ 1CD5	◌᳥ 1CE5	
6	◌᳖ 1CD6	◌᳦ 1CE6	
7	◌᳗ 1CD7	◌᳧ 1CE7	
8	◌᳘ 1CD8	◌᳨ 1CE8	
9	◌᳙ 1CD9	ᳩ 1CE9	
A	◌᳚ 1CDA	ᳪ 1CEA	
B	◌᳛ 1CDB	ᳫ 1CEB	
C	◌᳜ 1CDC	ᳬ 1CEC	
D	◌᳝ 1CDD	◌᳭ 1CED	
E	◌᳞ 1CDE	ᳮ 1CEE	
F	◌᳟ 1CDF	ᳯ 1CEF	

Devanāgarī Extended (U+A8E0 – U+A8FX)

	A8E	A8F
0	◌꣠ A8E0	◌꣰ A8F0
1	◌꣡ A8E1	◌꣱ A8F1
2	◌꣢ A8E2	ꣲ A8F2
3	◌꣣ A8E3	ꣳ A8F3
4	◌꣤ A8E4	ꣴ A8F4
5	◌꣥ A8E5	ꣵ A8F5
6	◌꣦ A8E6	ꣶ A8F6
7	◌꣧ A8E7	ꣷ A8F7
8	◌꣨ A8E8	꣸ A8F8
9	◌꣩ A8E9	꣹ A8F9
A	◌꣪ A8EA	꣺ A8FA
B	◌꣫ A8EB	ꣻ A8FB
C	◌꣬ A8EC	
D	◌꣭ A8ED	
E	◌꣮ A8EE	
F	◌꣯ A8EF	

Devanāgarī Block

ADDITIONS IN DEVANĀGARĪ BLOCK

0900	◌ऀ	DEVANAGARI SIGN INVERTED CANDRABINDU *=vaidika adhomukha candrabindu*
0955	◌ॕ	DEVANAGARI VOWEL SIGN CANDRA LONG E • *used in transliteration of Avestan*

OTHER CONSONANTS

0979	ॹ	DEVANAGARI LETTER ZHA • *used in transliteration of Avestan*
097A	ॺ	DEVANAGARI LETTER HEAVY YA • *used for an affricate glide JJYA*

PṚṢṬHAMĀTRA ORTHOGRAPHY VOWEL SIGNS

094E	◌ॎ	DEVANAGARI VOWEL SIGN PRISHTHAMATRA E *•Character has historic use only*

Vedic Extensions (U+1CD0 – U+1CFX)

TONE MARKS FOR THE SĀMAVEDA

UTC Code	Glyph	Name of character/ symbol
1CD0	◌᳐	VEDIC TONE KARSHANA *= vaidika saamasvara karshanna*
1CD1	◌᳑	VEDIC TONE SHARA *= vaidika svarita uurdhva shara*
1CD2	◌᳒	VEDIC TONE PRENKHA *= vaidika saamasvara prenkha* *• indicates vibrato*
1CD3	᳓	VEDIC SIGN NIHSHVASA *= vaidika saamagaana yogakaala*

SIGNS FOR YAJURVEDA

1CD4	◌᳔	VEDIC SIGN MID-CHARACTER SVARITA *= vaidika svarita urasi rekhaa*
1CD5	◌᳕	VEDIC TONE YAJURVEDIC AGGRAVATED INDEPENDENT SVARITA *= vaidika svarita adho nyubja*
1CD6	◌᳖	VEDIC TONE YAJURVEDIC INDEPENDENT SVARITA *= vaidika svarita adhah konna*
1CD7	◌᳗	VEDIC TONE YAJURVEDIC KATHAKA INDEPENDENT SVARITA *= vaidika svarita adho vakra rekhaa*
1CD8	◌᳘	VEDIC TONE CANDRA BELOW *= vaidika svarita adho ardha vakra*

1CD9	◌᳙	VEDIC TONE YAJURVEDIC KATHAKA INDEPENDENT SVARITA SCHROEDER *=vaidika svarita adho samyukta rekhaa*
1CDA	◌᳚	VEDIC TONE DOUBLE SVARITA *= vaidika svarita uurdhva dvi rekha*
1CDB	◌᳛	VEDIC TONE TRIPLE SVARITA *= vaidika svarita uurdhva tri rekhaa*
1CDC	◌᳜	VEDIC TONE KATHAKA ANUDATTA *= vaidika svarita adho rekhaa*
1CDD	◌᳝	VEDIC TONE DOT BELOW *= vaidika svarita adho bindu*

TONE MARKS FOR THE SATAPĀṬHABRĀHMAṆA

1CDE	◌᳞	VEDIC TONE TWO DOTS BELOW *= vaidika svarita adho dvi bindu*
1CDF	◌᳟	VEDIC TONE THREE DOTS BELOW *= vaidika svarita adho tri bindu*

Tone mark for the ṚGVEDA

1CE0	◌᳠	VEDIC TONE RIGVEDIC KASHMIRI INDEPENDENT SVARITA *=vaidika uurdhva vakra rekhaa*

TONE MARK FOR THE ATHARVAVEDA

1CE1	◌᳡	VEDIC TONE ATHARVAVEDIC INDEPENDENT SVARITA *= vaidika svarita dvi vakra khannda*

DIACRITICS FOR VISARGA

1CE2	◌᳢	VEDIC SIGN VISARGA SVARITA *=vaidika madhyarekhaa*
1CE3	◌᳣	VEDIC SIGN VISARGA UDATTA *= vaidika visarga dakshinnatah uurdhvaga*
1CE4	◌᳤	VEDIC SIGN REVERSED VISARGA UDATTA *= vaidika visarga vaamatah uurdhvaga*
1CE5	◌᳥	VEDIC SIGN VISARGA ANUDATTA *= vaidika visarga vaamatah adhoga*
1CE6	◌᳦	VEDIC SIGN REVERSED VISARGA ANUDATTA *= vaidika visarga dakshinnatah adhoga*
1CE7	◌᳧	VEDIC SIGN VISARGA UDATTA WITH TAIL *= vaidika visarga dakshinnatah uurdhva vakra*
1CE8	◌᳨	VEDIC SIGN VISARGA ANUDATTA WITH TAIL *= vaidika visarga vaamatah adho vakra*

MARKS OF NASALISATION

1CE9	ᳩ	VEDIC SIGN ANUSVARA ANTARGOMUKHA = *vaidika anusvaara antarmukha*
1CEA	ᳪ	VEDIC SIGN ANUSVARA BAHIRGOMUKHA = *vaidika anusvaara naagaphanna*
1CEB	ᳫ	VEDIC SIGN ANUSVARA VAMAGOMUKHA = *vaidika anusvaara vaamamukha*
1CEC	ᳬ	VEDIC SIGN ANUSVARA VAMARGOMUKHA WITH TAIL = *vaidika anusvaara vaamagomukha sa-vakra*
1CED	◌᳭	VEDIC SIGN TIRYAK =*vaidika tiryak*
1CEE	ᳮ	VEDIC SIGN HEXIFORM LONG ANUSVARA =*vaidika anusvaara anugamii*
1CEF	ᳯ	VEDIC SIGN LONG ANUSVARA =*vaidika anusvaara dakshinnamukha*
1CF0	ᳰ	VEDIC SIGN RTHANG LONG ANUSVARA =*vaidika anusvaara ttha-sadrisha*

1CF1	ᳱ	VEDIC SIGN ANUSVARA UBHAYATOMUKHA =*vaidika anusvaara ubhayatomukha*

ARDHAVISARGA

1CF2	◌ᳲ	VEDIC SIGN ARDHAVISARGA =*vaidika jihvaamuliiya upadhamaaniiya*

Devanāgarī Extensions (U+A8E0 – U+A8FX)

CANTILLATION MARKS FOR THE SĀMAVEDA

A8E0	◌꣠	COMBINING DEVANAGARI DIGIT ZERO =*vaidika saamasvara anka shuunya*
A8E1	◌꣡	COMBINING DEVANAGARI DIGIT ONE =*vaidika saamasvara anka eka udaatta*
A8E2	◌꣢	COMBINING DEVANAGARI DIGIT TWO =*vaidika saamasvara anka dvi svarita*
A8E3	◌꣣	COMBINING DEVANAGARI DIGIT THREE =*vaidika saamasvara anka tri anudaatta*
A8E4	◌꣤	COMBINING DEVANAGARI DIGIT FOUR =*vaidika saamasvara anka chatur*
A8E5	◌꣥	COMBINING DEVANAGARI DIGIT FIVE =*vaidika saamasvara anka pancha*
A8E6	◌꣦	COMBINING DEVANAGARI DIGIT SIX =*vaidika saamasvara anka shatt*
A8E7	◌꣧	COMBINING DEVANAGARI DIGIT SEVEN =*vaidika saamasvara anka sapta*
A8E8	◌꣨	COMBINING DEVANAGARI DIGIT EIGHT =*vaidika saamasvara anka ashta*

A8E9	◌꣩	COMBINING DEVANAGARI DIGIT NINE =*vaidika saamasvara anka-nava*
A8EA	◌꣪	COMBINING DEVANAGARI LETTER A =*vaidika saamasvara abhinihita*
A8EB	◌꣫	COMBINING DEVANAGARI LETTER U =*vaidika saamasvara u*
A8EC	◌꣬	COMBINING DEVANAGARI LETTER KA =*vaidika saamasvara ka*
A8ED	◌꣭	COMBINING DEVANAGARI LETTER NA = *vaidika saamasvara namana*
A8EE	◌꣮	COMBINING DEVANAGARI LETTER PA =*vaidika saamasvara prannatam*
A8EF	◌꣯	COMBINING DEVANAGARI LETTER RA =*vaidika saamasvara ra*
A8F0	◌꣰	COMBINING DEVANAGARI LETTER VI =*vaidika saamasvara vinata*
A8F1	◌꣱	COMBINING DEVANAGARI SIGN AVAGRAHA =*vaidika saamasvara diirghiibhaava*

MARKS OF NASALISATION

A8F2	ꣲ	DEVANAGARI SIGN SPACING CANDRABINDU =*vaidika candrabindu*
A8F3	ꣳ	DEVANAGARI SIGN CANDRABINDU VIRAMA =*vaidika anusvaara candrabindu tiryak*
A8F4	ꣴ	DEVANAGARI SIGN DOUBLE CANDRABINDU VIRAMA =*vaidika anusvaara dvi candrabindu tiryak*
A8F5	ꣵ	DEVANAGARI SIGN CANDRABINDU TWO =*vaidika anusvaara candrabindu sa-dvi*
A8F6	ꣶ	DEVANAGARI SIGN CANDRABINDU THREE =*vaidika anusvaara candrabindu sa-tri*
A8F7	ꣷ	DEVANAGARI SIGN CANDRABINDU AVAGRAHA =*vaidika anusvaara candrabindu sa-avagraha*
A8F8	꣸	DEVANAGARI SIGN PUSHPIKA =*vaidika pushpika* • used as a placeholder or "filler" • often flanked by double dandas

A8F9	꣹	DEVANAGARI GAP FILLER used to indicate that a gap in a manuscript or text is not a lacuna
A8FA	꣺	DEVANAGARI CARET =*vaidika trutikaa* • zero-width character centred on the point between two orthographic syllables
A8FB	ꣻ	DEVANAGARI HEADSTROKE =*shirorekhaa* used to indicate uncertain manuscript reading

The UTC recommended proposal was further discussed in the 53rd Meeting of the ISO Working Group-2 "Universal Coded Set (UCS)" of JTC1/SC2 in Hong Kong, China, during October 13-17, 2008. Mrs. Swaran Lata, Director and HOD, Human Centred Computing Division, DIT, participated in the meeting as the representative of Government of India and presented/discussed the proposal. The proposal has now been accepted by ISO also and after the April 2009 meeting of the above working group of ISO, the code-set will be included in the future version of ISO10646 and Unicode Standard.

The charts shown in the previous pages explain the various characters accepted for encoding for the representation of Vaidika characters and symbols as recommended by Department of Information Technology, Ministry of Communications and IT, Government of India, for inclusion in the future version of ISO10646 and Unicode Standard.

Endnotes

1. Vishwabharat@tdil: *Language Technology Flash* Volume, October 2002.
2. www.unicode.org

18

Sanskrit, Whole Brain Learning
And its Importance in the Super Accelerated Learning Theory (SALT)

Pandit Ramadheen Ramsamooj

Abstract

Sanskrit is one of the oldest languages known to man developed by Ṛṣi Pāṇini and then Patañjali who was a profound Yoga master, well versed in language structure and form. Sanskrit is a language that has unique features in its phonetics, script, and sentence structure. The sound and form in its alphabet and Devanāgarī script form and structure allows for the stimulation of the whole brain in the processing of information and memory. This language was designed for efficient communications both at the conscious and supraconscious levels. It conveys knowledge at both the material and spiritual levels and is used as a vehicle for intuitive learning and universal knowledge transfer.

Introduction

SANSKRIT is known as a whole brain language because of its form and structure. It first begins with sounds and how they are processed by the brain. Static sounds, viewed as noise, are processed by the left hemisphere of the brain while sonorous sounds are processed by the right hemisphere of the brain. Noise or static sound is a left brain phenomenon while musical sonorous sounds are a phenomenon of the right brain. Many people based on their cultural background and brain development view sounds very differently. In some culture all sounds are interpreted as noise and as a result it is processed by the left brain. In other cultures, sounds like the rustling of the leaves, rippling of the waters in a stream, chirping of the birds, and the hissing of the wind, sounds of nature are sounds that instill a sense of peace and calm are processed by the right brain and thus are never viewed as noise. Sanskrit as a language is looked at several levels of its construct in establishing it as a whole brain language using some of the research done on brain theory.

The Super Accelerated Learning Theory (SALT) is based on learning using the whole brain in the processing of information. Sanskrit as a language helps in the wiring of both hemispheres of the brain making it ready for rapid learning and processing of information.

This paper looks at the features and form of Sanskrit, whole brain theory, the SALT and the role Sanskrit plays in the stimulation and development of various neurological pathways in the different hemispheres of the brain and in the wiring of the brain for rapid learning.

The Brain and its Hemispheres

The human brain has two hemispheres, a left and a right, and is connected by the hypothalamus. It also has the neocortex for processing complex algorithm, a mammalian brain that governs instincts and the reptilian brain that response to fright and flight. This has been a succinct description of the evolution of the brain and its development with its various functions.

From brain research and its configuration it is quite clear that the left and right hemispheres of the brain have different functions and are stimulated differently. The left hemisphere is associated with speech, reading, calculating and deducing, reason, ratio and logic, rules and law, grammar, fine concentration, focus, analysis, detail, science, temporal sense,

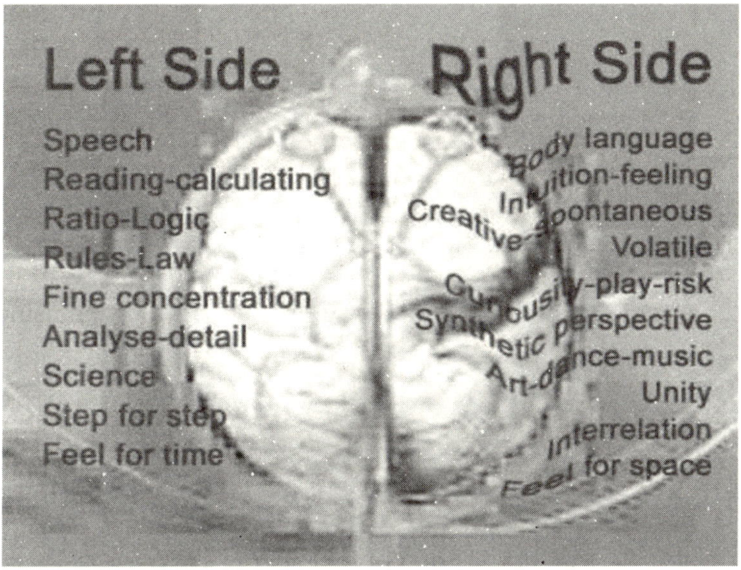

metre and rhythm. The right hemisphere is associated with body language — signs and symbols, intuition, feeling, creativity, spontaneous and volatile curiosity, play, risk, synthetic perspective, art, dance, music, unity, interrelation and sense of space. All of these functions and associations are depicted in the Sanskrit language.

Sanskrit is looked upon at several levels in respect to whole brain theory and learning.

Level 1: Sanskrit Alphabet Structure

PHONETICS

Sanskrit has categorised its alphabet in two separate groups, the sonants and consonants. The sonant or *svara* (vowels) consists of 16 different sounds, one short and the other long. All sounds come from the throat region which vibrates along the spinal cord and thus has its subtle influence on the brain. Short sounds, such as अ, इ, उ, ऋ, लृ, ए, ओ, अँ are perceived and interpreted by the left hemisphere of the brain, while the long sounds, more musical in nature, आ, ई, ऊ, ॠ, लॄ, ऐ, औ, अः are perceived and interpreted by the right hemisphere of the brain. In the language construct, the vowel sounds inherently have left and right brain stimulation — left for the short vowels and right for the long vowels due to its sonorous feature. These, when combined with the consonants, make Sanskrit a whole brain stimulated language. It should be noted that 16 vowels in the Devanāgarī alphabet is very unique to this language. The number 16 with its rhythm plays an important role in information storage and retrieval.

SANSKRIT ALPHABET STRUCTURE

(a.) Vowel sounds (16 Vowels)

अ	आ	इ	ई
उ	ऊ	ऋ	ॠ
ऌ	लॄ	ए	ऐ
ओ	औ	अँ	अः

The *svara* (vowel) sounds are placed in a 4 × 4 matrix. In this form, the brain looks at a whole picture with these 16 letters. Arranged in this manner, the brain interprets this array as a picture and places this information in the right hemisphere of the brain. With the different sounds short and then long places the sound in the left hemisphere and right hemisphere. With the visual array and the sound new pathways are formed with the letters and the sounds and engages the whole brain. With the vowel system alone, the whole brain is being engaged and neurons are being fired in both hemispheres.

b.Consonants : Phonetics, Form and shapes

क ख ग घ ङ Velar

च छ ज झ ञ Palatal

ट ठ ड ढ ण Retroflex

त थ द ध न	Dental
प फ ब भ म	Labial
य र ल व	
श ष स ह	
क्ष त्र ज्ञ	

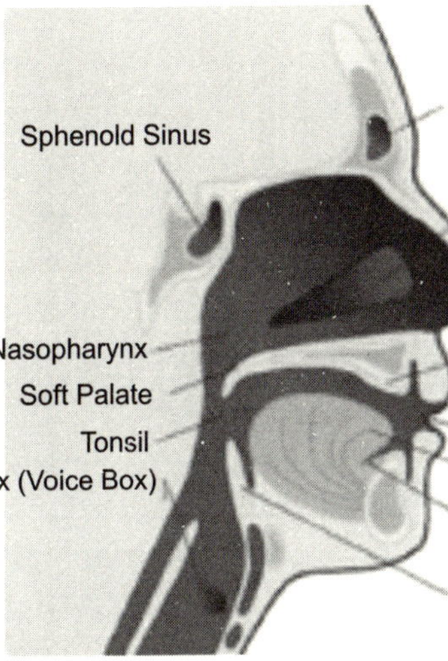

Points of Articulation

The production, of the sounds of the consonants is very ordered, following a very sequential pattern in the mouth. The first row of sounds is produced in the base of the throat and the sounds are classified as velar. The second row is from the palate region. The third from the hard palate region called *mūrdhā*, the fourth the dental and the fifth is from the lips — the labial. This order and sequence logically produced is processed by the left brain. The sound production properly done not only produces an audible sound but also produces an inaudible vibration that creates a different set of neurological pathways in the brain and along the spinal cord. This leads to information being extended beyond the physical brain. Different parts of the mouth are associated with different parts of the brain. Thus different sounds are placed in different positions in both the hemispheres of the brain. Phonetically the use of the right and left ears as channels of hearing the various sounds, conveyed are stored in different sites and pathways in the brain.

FORM AND SHAPE

As the consonants are displayed above, the brain views the whole array of consonants as one picture and stores it in one site in the right hemisphere of the brain. Forms and shapes are registered differently in the brain. As individual letters, they are stored in different places thus, establishing various physical sites for the storage of the same information. Lines and their variations are perceived as representatives of numbers and are processed in the left hemisphere of the brain. Shapes that are viewed as images or pictures are processed in the right hemisphere of the brain. If we look at the shape and form of the consonants we can see a definite pattern of line, shape, letters and image or picture-shaped letters in the Devanāgarī script. If we look at the velar sounds

क line ख picture ग line घ picture

क is viewed as a line form. It consists of three lines — horizontal, vertical and S curve line with a 90 degree rotation put together forms **क.**

Thus, क is viewed as a line. The same can be said for ग, etc.

The second letter ख is viewed as a picture because of the complex curves and lines, similarly the same can be said for घ, छ, etc.

The consonants can then be viewed as follows:

क line ख picture ग line घ picture ङ	Velar
च line छ picture ज line झ picture ञ	Palatal
ट line ठ picture ड line ढ picture ण	Retroflex

त line थ picture द line ध picture न Dental

प line फ picture ब line भ picture म Labial

The first column and the third column are viewed as lines processed by the left hemisphere of the brain and the second and fourth columns are viewed as pictures processed by the right hemisphere of the brain. In the learning of the alphabet there is a constant shift from left hemisphere to right hemisphere in the processing of the information. This exercise helps to wire the neurons in the brain with crossover between the left and right hemispheres and right and left hemispheres. This is different from neurons connecting only in the left hemisphere and neurons being connecting only in the right hemisphere. This is what makes the difference between left brain learning, right brain learning and whole brain learning.

Another aspect of the alphabet is that the sound of the letters and the form and shape of the letters *simultaneously* excite the neurons to fire in all directions in the whole brain, both right and left hemispheres and through the hypothalamus. This prepares the brain to rapidly process information.

Level 2: Sound of Words with Visuals

In Level 1 we looked at the alphabet construct and all the possible sounds in the vowels and consonants and their possible combinations. In Level 2, we look at the word constructions and the ideas that they convey. Productive nature of the language which derives each word from a limited set of roots by way of derivation processes, has the capacity of describing complicated concepts in fewer words. As a result, a small compact text in *sūtra* can convey very complicated concepts. Pāṇini's grammar has only 4000 statements and can describe entire Sanskrit language. This tradition has been very prevalent in the Sanskrit language. As a language that conveys meaning in a different plane of understanding it is very important to present those concepts in simple terms. Difficult concepts from other languages too can be easily rendered in Sanskrit using its inherent capacity precise constructions.

Level 3: Structure and Sentences

Sanskrit is a relatively free word-order language due to its being highly inflected. The words have their own declension according to case and number and independently have their complete meaning. This is very valuable in efficiency and in the processing of information. Words can be placed at random in both hemispheres creating a variety of pathways for processing. This is also allowed for stacking of information in one physical site and therefore allows it to be recalled instantly. With the sonorous *mantra* recital, connections are generated at different stacking levels, generating a network of information even beyond the conscious level. This process stimulates the whole brain in information storage and retrieval. Forty thousand verses from the Vedas can be easily remembered by a young child. There are also so many independent pathways created that one can perform several tasks at the same time. *a*· Ashtavadhan's, Shathavadhan's and mathematical genius such as Shakuntala Devi — the walking human computer can be made possible. It is this feature of the language that makes it a natural computer programming language.

Level 4 : At the Intuitive Level

With the constant resonance of the sounds, *mantras*, *sūktas* and proverbs, the brain is constantly processing information in both hemispheres. This leads to the constant activation of the right hemisphere that is known for the expression of the intuitive ability. Intuition is the ability to predict an outcome without going through the logical steps. This is a result of parallel processing. It is this ability that allows mothers to go with their gut instinct in making their decision.

Level 5: Yoga Level Disciplined Mind

In disciplining the mind small profound thoughts are embedded for reflection in quieter times. The *Yoga-Sūtras* — strings of information, precise and concise reflecting practice and control — are used as the foundation for disciplining the mind for students wishing to study spiritual knowledge. There are also the *sūkta*s, derived from the *Ṛgveda*, proverbial thoughts that one reflects lifelong. Even physical postures or *āsana*s can bring about a tremendous amount of coordination between physical and mental activities. *Haṭha-yoga* with its breathing and different body orientation stimulates the whole brain.

Level 6: Learning in an Altered State

Information can flow as a pulse, squirting a little at a time. This is done in a normal state of mind. Information can also flow in a wave motion which is a certain oscillation. This is experienced in a semi-altered state of consciousness. In the altered state of consciousness the information flows as a broad band of light connecting every site of information whether it is stacked or not. This leads to very rapid flow and processing of information. Here, connection is made at the cosmic conscious level. In this state new knowledge is revealed. Sanskrit through its rhythmic *mantra* recitation put one in a meditative state and then lift their consciousness to an elevated state and ultimately take them in the information super highway.

Level 7: Language that is Emotionally Uplifting

Sanskrit is not the language that is emotionally degenerating. It is structured with words to lift the spirit and emotion at a level of consciousness that is beyond that of human experience. There are certain words and concepts that can only be found in Sanskrit. Words such as *ānanda, dharma, karma, mokṣa, yoga, prāṇāyāma, ātman, parmātman, oṁ* are not found in other languages. Actually other languages have borrowed these words in their own vocabulary. In English we have *omnipresent* (all present), *omnipotent* (all powerful) and *omniscient* (all knowledgeable). The idea of wholeness, oneness, in its totality, all-encompassing can only be conveyed in the word *oṁ*.

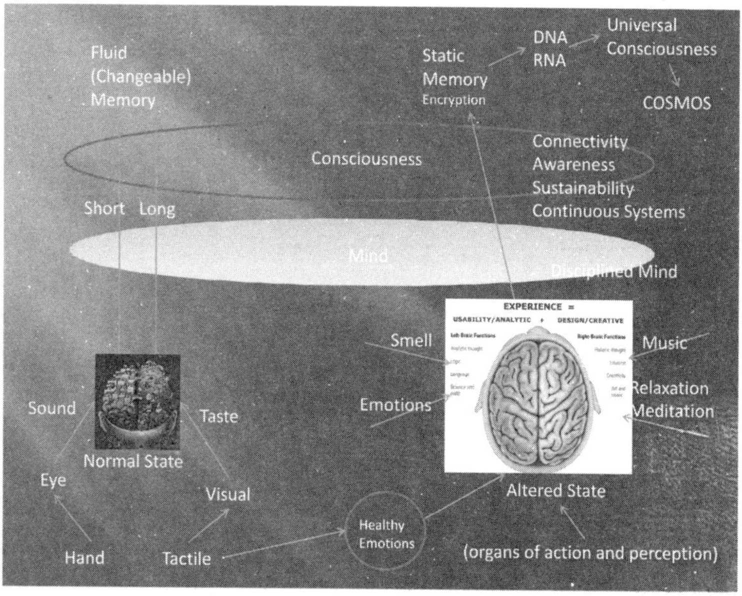

The vast amount of literature that is conveyed in this language is phenomenal. It begins with the Vedas, the Brāhmaṇas, Āraṇyakas, Upaniṣads, Kalpasūtras, Vedāṅgas, Purāṇas and epics. This whole body of knowledge is designed to uplift the spirit and soul of the being. Much of it is done through storytelling, the most effect way of conveying ideas and thoughts.

Method of Teaching Sanskrit

In the teaching of Sanskrit, it is important that the tradition is kept intact. This method involves memorisation, being at the feet of the *guru*, language embedded in Yoga, language portraying profound life philosophy, related to wholeness and non-attachment, mental discipline and focus, different sounds being listened by different ears, left ear right brain, right ear left brain. In this teaching method the teacher is important. He is the knower and represents the supreme source. Knowledge is communicated from the teacher to the student. In the *śruti vidyā* learning tradition, the *guru* plays a very important role. The *uccāraṇa* (recitation) of *mantra*s must be perfect and the student must be able to remember it at all times. For this the *guru* must prepare the student to learn. For this there is an initiation ceremony.

Initiation Ceremony

In the initiation ceremony, the student is first introduced to sound, then form letters and words, then meaning to context, then to abstract concepts ultimately to the cosmic wholeness. Apart from the constant recitation of *mantra*s and performance of rituals, the *guru* gives *dīkṣā* by giving the *guru-mantra* both in the spoken and written forms. In this ceremony, the student sits on the right lap of the teacher with a book or container with rice on the right side and the *guru* speaks in the left ear. He holds the pen or finger and moves the student's hands to form the letter. The student's eyes are to the right, stimulating the left hemisphere and the sound in the left ear stimulating the right hemisphere. Through sight and sound, both hemispheres are activated simultaneously. Constant repetition of the *mantra* leads to an altered state of consciousness. Through memorisation, information is placed first in the right hemisphere. The meaning is obtained at a later time by reason and deduction. Memorisation puts information in the right brain first then later finds meaning and analysis.

Brain and its Functions

From brain research, it has been found that different parts of the brain have different functions. Dr. Howard Gardner postulated his theory of multiple intelligences on this very aspect of the brain. However, at the time not much research was done on the sites and function of the brain. The whole brain theory of learning takes into consideration all functions of the brain.

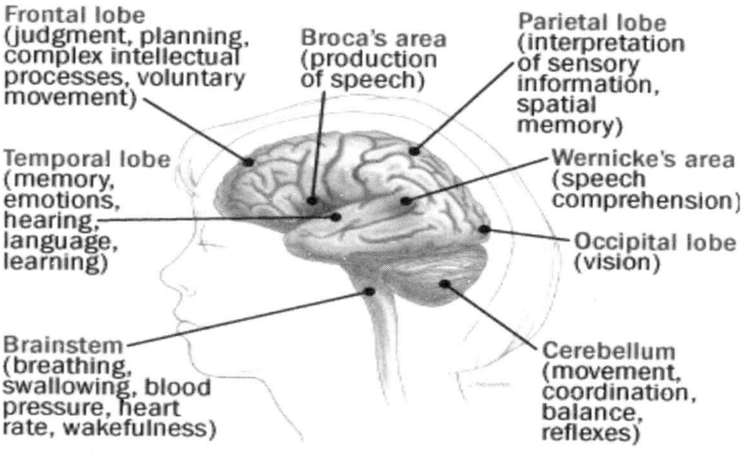

Diagram Showing Brain and Its Functions

For Super Accelerated Learning the brain must be wired in a particular way for rapid processing of information. Linear and parallel processing are the architecture of the left and right brain. The SALT is based on the whole brain development of which Sanskrit plays an important role.

Super Accelerated Learning Theory

The SALT states that the acquisition, assimilation and processing of Information of various disciplines, holistically, are through all organs of actions and perceptions, in the information super highway, with the use of the whole brain and a disciplined mind, in a semi or altered state of consciousness. It is based on the psychological development and physiological function of the biological brain, through various neurological pathways, in respect to learning, application, storage and retrieving of information, in and from its long-term static memory, instantaneously.

In SALT there is a unique way in which information is presented. On analysis, there are many similarities in the way Sanskrit is taught and the way information is presented in SALT.

SALT: Presentation of Information

- Story format
- Emotionaly powered, drama packed
- With different connectors, different pathways
- From whole to parts, then parts to whole, global to local
- Connectivity and purpose-driven
- Use colours and visuals wherever necessary
- Use poetry, rhymes and music
- Quick mental review, closed eyes, relaxed state
- Rapid question and answers
- Create a game with information, make learning fun

Multiple Intelligence Development

According to the Vedic tradition, learning takes place before birth. They have designed rituals called *saṁskāras* that are performed before birth. These rituals of impressions are designed to stimulate the mind and the thinking faculty of the child during gestation. Dr. Gardner has spoken of seven intelligences (SKILLMI) Spatial and Space Orientation, Kinesthetic, Interpersonal, Logics and Mathematics, Language, Music and Intrapersonal. I have introduced two more intelligences — Self-Restraint and Spiritual Affinity.

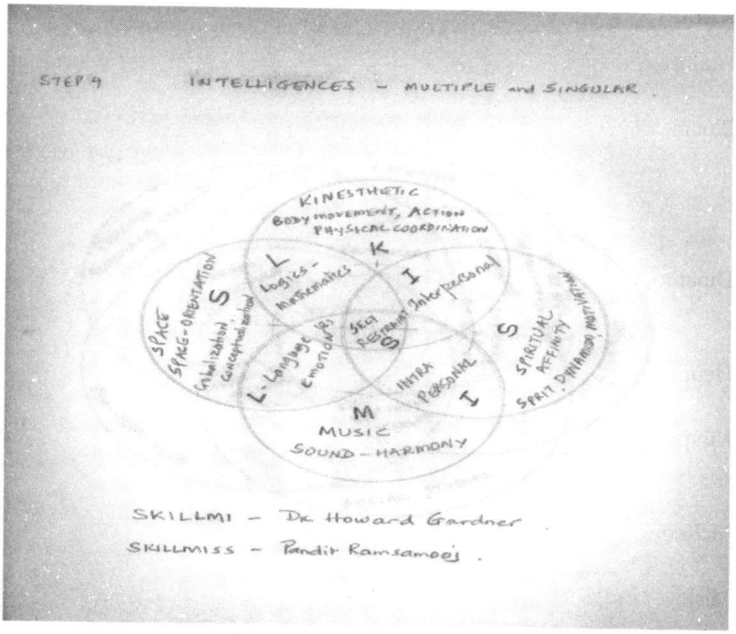

Sanskrit through its recitation of *mantras* forces you to breathe properly. Breath involving the *iḍā* and *piṅgalā* has an alternating effect on the brain function. Kālidāsa's epic, musicals, poetry, *ślokas*, *mantras*, *sūtras* all help in developing the multiple intelligences in the child. Dance rhythm based on the Sanskrit dramas gave Bharatanāṭyam, Kathaka and other dance forms that have intricate feet movement, hand movements, eyes movement conservation of space and space orientation making it a whole brain discipline. Sanskrit is known as the language of the Gods. It invokes subtly human and ethical values and lifts one's consciousness to a divine existence, leaving the students with deep moral, ethical and spiritual values, a prerequisite for a good education, a SALT education.

References

Atkinson, Alan, "An Interview with Libyan Labiosa Cassone", *The Learning Revolution*, Winter 1991, P.22

Bloom, Howard, Sandra Ham; Susie Kagehiro; Laura Melton; Julieanne O'Brien; JoAnn Rock; and Doolittle Fred, *Evaluating the Accelerated Schools Program: A Look at Its Early Implementation and Impact on Student Achievement in Eight Schools,* New York: Manpower Development Research Corporation 2000.

Dipamo, B. and R.F.S. Job, "A Methodological Review of Studies of SALT (Suggestive-Accelerative Learning and Teaching) Techniques", *Australian Journal of Education Technology* 1991, 7(2), 127-43.

Heppr, P. "The Musical Foetus", *Irish Journal of Psychology* 1991, 12, pp. 95-107.

Kuyper-Erland, Jan, *Brain-Based Accelerated Learning Longitudinal Study Reveals Solid Academic Achievement Maintenance With Accelerated Learning Practice.*

McKeon, K. J., "What Is This Thing Called Accelerated Learning", *Training and Development* 49, no. 6 (June 1995): 64-66.

Mislevy, R.J., L.S. Steinberg, R.G. Almond, G.D. Haertel and W.R. Penuel (2003), "Improving Educational Assessment", in B. Means and G.D. Haertel (eds.), *Evaluating Educational Technology: Effective Research Designs for Improving Learning,* pp. 149–80. New York: Teachers College Press.

Web References

www.singsurf.org/brain/rightbrain.php, *Left Brain, Right Brain,* Whole Brain?

www.ucop.edu/ucophome/businit/boi/presentations/2005/05styles.pdf, *The Whole Brain Model: Understanding Working Styles*

www.americansanskrit.com/read/a_techage.php, *Sanskrit and the Technological Age*

http://unzipyourgenes.files.wordpress.com/2011/04/human-blockhead-nasal-diagram.gif

Contributors

Nabanarayan Bandyopadhyay professor and director, School of Vedic Studies Rabindra Bharati University, Kolkata, holds a brilliant academic record and has teaching and research experience of 31 years. He has visited many countries abroad and participated in a number of state level, nantional, international seminars, symposia, worshops, conferences, refresher courses, etc. and presented learned papers. There are a lot of published research papers and books to his credit. Some of the edited books include *Ancient Indian Views on Truth and Falsity, Mīmāṁsā-Paribhāṣā, Laghu-Yajñapārśva-Pariśiṣṭa and Mahad-Yajñapārśiṣṭa, Bengal's Contribution to Vedic Studies, Vedic Studies*: vols. III, IV, V, etc. He has successfully supervised candidates for earning their Ph.D degrees from the University. He has also regularly conducted/participated in various major research projects. A project on Vedic Ancillary Texts on Phonetics has recently been undertaken by the School under his leadership. [*address*: Professor and Director of the School of Vedic Studies, Rabindra Bharati University, Kolkata, West Bengal, *e-mail*: <naba_narayan@yahoo.com>].

Soma Basu, has had a very good academic carrer (1st class in all the examinations passed) upto her Ph.D, and worked for various research projects in Rabindra Bharati University, Kolkata; the Asiatic Society, Kolkata; and K.J. Somaiya Centre for Indological and Buddhist Studies, Mumbai; and as a lecturer for the UGC major project on the Vedas and Vedic Sciences in the School of Vedic Studies. Apart from her proficiency in Sanskrit, she has ample knowledge of Russian language (certificate course in Russian language from the Kolkata branch of All India Institute for Russian Language, New Delhi) and computer operations (from YMCA, Kolkata). Her research papers have been published by various institutions in Kolkata, Mumbai, Sarnath, New Delhi, etc. She has participated in a number of national and international seminars, conferences and workshops, and presented papers relating to Buddhistic studies, Vedic studies, ancient Indian history, rites and rituals, etc. Her monograph entitled *Supriya Sārthavāha Jātaka of Bhadrakalpāvadāna* has been published by the Asiatic Society, Kolkata in the Bibliotheka Indica Series 335 in 2010 [*address*: Associate Professor, School of Vedic Studies, Rabindra Bharati University, Kolkata, *e-mail*: <somabasu01@gmail.com>].

Sadasiva Biswal, a retired principal, born in 1945 in Orissa; got his Ph.D degree from Utkal University, Bhubaneswar. He worked as the reader and head of the department of Physics of several colleges in Orissa and has produced five Ph.Ds. He has published thirty-five research papers and more than two hundred articles on science besides 14 books on Physics. Dr. Biswal is associated with many scientific and research organisations in India and abroad, and is in the Research Board of Advisers of the American Biographical Institute Inc., Raleigh, North Carolina, USA. [*address*: Plot no. - 193, Jayadev Vihar, Near SBI ATM, Bhubaneswar - 751 013; *e-mail*: <dr.sadasivabiswal@ymail.com>].

Manoj Jain, a Bachelor of Engineering in Electronics and M.Tech. in Digital Communications Engineering from Bhopal University, is presently working in the "Technology Development for Indian Languages" programme of the Department of Information Technology, as Scientist "E". He had a key role in the Unicode representation of Indian languages. He is also looking after the development of Machine Translation Systems for Indian languages and various electronic resources for Indian languages. [*address*: Scientist E, TDIL Programme, Department of IT, Electronics Niketan, 6, CGO Complex, Lodhi Road, New Delhi, *e-mail*: <mjain@mit.gov.in>].

Dillip Kumar Kar was born in 1970 in Nayagarh district of Odisha state and completed his Acharya (M.A.) and Vishisthacharya (M.Phil.) degree in Advaita Vedanta Philosophy from Sri Jagannatha Sanskrit Vishvavidyalaya, Puri in 1993 and 1995 respectively and was awarded as Vidyavaridhi (Ph.D) by the Rashtriya Sanskrit

Sansthan, New Delhi in 2000. He joined as a research assistant in the Academy of Sanskrit Research, Melkote, Karnataka in 1999 and worked on different Oriental Science projects such as — (i) "Iron and Steel in Ancient India: A Study in Historical Perspective" sponsored by the Steel Authority of India, (ii) "A non-conventional approach to Indian Aeronautics" and (iii) "Science and Technology in India through the Ages" sponsored by ISRO, Bangalore. In 2005 he left the job from the Academy and joined in the National Mission for Manuscripts, New Delhi. At present he is holding the post of co-ordinator (Survey) of the Mission [*address*: National Mission for Manuscripts, IGNCA, 11 Mansingh Road, New Delhi - 110 001, *e-mail*: <prachyaprajna@gmail.com>].

Shashiprabha Kumar, former chairperson of the Special Centre for Sanskrit Studies, JNU, is an acclaimed Sanskrit scholar, well-versed in classical Indian philosophy. Before joining JNU, she taught at the department of philosophy, University of Delhi; department of Sanskrit at Maitreyi College, New Delhi and R.G. Postgraduate College, Meerut. More than twenty research scholars have obtained M.Phil/Ph.D. under her guidance. She has fourteen books to her credit and has contributed more than seventy research papers to reputed journals and edited several volumes. She has participated in many International Conferences and National Seminars besides lecturing in India and abroad. Recipient of Sri Ramakrishna Sanskrit Award (World Education Foundation, Canada, 2003) for her distinguished and outstanding contribution to Sanskrit research and teaching and **Shankar Puraskar** (K.K. Birla Foundation, New Delhi, 1998) for her first book entitled *Vaiśeṣika Darśana mein Padārtha-Nirūpaṇa*, she has won many other prestigious awards and fellowships. [*address*: *Abhyudaya*, 295/Sector - 15-A, Noida (U.P.) - 201 301, *e-mail*: <shashiprabhakumar@yahoo.co.in>].

Swaran Lata, is heading the department of Information Technology's National Strategic Programme "Technology Development for Indian Languages", programme of the Department of Information Technology, Govt of India. She is also country manager, W3C, India Office. She holds bachelor's degree in Electronics and Communications, with Gold Medal, from the Thapar Inst. of Engg.

and Technology and M.Tech in Computer Technology from IIT Delhi. Mrs. Lata is also member of various international standardisation bodies such as, ISO, UNICODE, W3C and ELRA and plays a leading role in finalising the UNICODE encoding system for 22 Indian languages and Vedic Sanskrit and establishing W3C India Office. She played a key role in initiating mission mode consortium projects in the area of Machine Translation Systems, Optical Character Recognition System, Cross Lingual Information Access System and Speech Technologies. [*address*: Department of Information Technology, Ministry of Communications and Information Technology, Electronic Niketan, 6, C.G.O. Complex New Delhi-110 003, *e-mail*: <slata@mit.gov.in>].

C.S.R Prabhu obtained his Masters Degree in Electrical Engineering with specialisation in Computer Science from IIT, Bombay. He remained head of various institutes of repute for a specific period of time such as the National Data Centre, Hyderabad (2007-08); Computer Society of India, (2002-04); University of Central Florida (1981-82), etc. He has also authored several books such as: *Semantic Database Systems: A Functional Introduction; e-Governance — Concepts and Case Studies; Grid and Cluster Computing*, etc. Prabhu has received awards of repute like: Oracle e-Governance Excellence Award, 2006, for "innovation with IT" and "e-Panchayat"; CIO100 Award (2010); and Manthan Festival Award, 2010, etc. [*address*: Deputy Director General, National Informatics Centre, A-Block, BRKR G.O. Complex, Tank Bunk Road, Hyderabad - 500 063, *e-mail*: <ddg@ap.nic.in>].

Chandrabhan Prajapati, graduated from M.I.T.S. Gwalior, India with a degree in "Mechanical Engineering" in 2007, completed his post-graduation from IIT Delhi in "Design Engineering" in 2010. Currently he is working as an assistant professor in an engineering college in Lucknow since June 2010. [*address:* C/o Jairam Prajapati, Near Dak Bunglow, Ram Leela Ground, Gandhi Nagar, Mahoba - 210 427 (U.P.); *e-mail*: <chandrapra2007@gmail.com>].

M.G. Prasad is a professor of mechanical engineering at Stevens Institute of Technology, Hoboken, New Jersey. He is a Fellow of the American Society of Mechanical Engineers, a Fellow of the Acoustical Society of America and a Fellow of the Acoustical Society of India.

He has more than 100 publications in the area of acoustics and vibrations to his credit and has written four books on Hinduism [*address*: Professor, Department of Mechanical Engineering, Stevens Institute of Technology, Castle Point on the Hudson, Hoboken, New Jersey 07030; *e-mail*: <mprasad@stevens.edu>].

Premendra Priyadarshi, MBBS, MD, FRCP, a consultant physician, received his higher qualification in Medicine in UK, where he served in the NHS for five years before returning to India. He is a Fellow of the Royal College of Physicians, Edinburgh. He has written three books on Indology, viz. *India's Contributions to the West* (2004), *Zero is Not the Only Story* (2007) and *The First Civilization of the World* (2011). Dr Priyadarshi has lectured on topics like Role of India in Human Evolution (DNA studies), Ancient Indian Sciences and Caste System to conferences [*address*: B-41, National Appartments, Plot - 4, Sector - 3, Dwarka, New Delhi - 110 078; *e-mail*: <priyadarshi101@hotmail.com>].

Vaidyaratnam **R. Raghavan** started pursuing Ayurveda at the age of 12, at Rishikesh in the Himalayas that was followed by a formal allopathic degree. He has successfully treated hundreds of thousands of patients through Ayurvedic treatment since 1966 suffering from all known physical afflictions, from hyphertension to chronic hepatitis. A disciple of Swami Dathathreya, pulse diagnosis is the key element in his natural, holistic treatment for overcoming bodily imbalances. As president, International Foundation for Ayurvedic Research and Training Trust, Dr. Raghavan is spearheading a worldwide awareness campaign for promotion of Ayurveda [*address*: South Palace, Puthiyadom, Kanjoor PO, Ernakulam District Kerala ; *e-mail*: <drraghavan1@gmail.com>].

Pandit Ramadheen Ramsamooj has a graduate degree in robotic engineering from the University of Wisconsin and an M.Ed. degree from the Cambridge College, Massachusetts. He is a lecturer at the University of Massachusetts, Dartmouth. As a founding director of 3 R Foundation, he has been developing Super Accelerated Learning Theory (SALT) education model, which he is implementing at the American Vivekananda Academy in Hudson, NH. Elements if his SALT education model are being adopted in schools and universities in the US, Trinidad, and India. [*address*: Saraswati Mandiram Trinidad & Tobago Inc., American Vivekananda Academy, 181 Ramsingh Trace, Rochard Douglas Road, Barrackpore, Penal, Trinidad and Tobago (West Indies), *e-mail*: <swamijiusa@gmail.com>].

Ranajit Kumar Saha, has had his PhD in Optical and Magnetic Properties of Solid; taught in the undergraduate colleges as head of the Department of Physics, before taking over important administrative posts such as principal of Engineering College; deputy controller of examination, Calcutta University; conroller of examinations, Bengal Engineering College (deemed university), and Vidyasagar University; and as chairman, West Bengal School Service Commission.

he has four renowned publications to his credit and presented research papers in the All India Oriental Conference at Varanasi and National Seminar organized by the School of Vedic Studies, Rabindra Bharati University, Kolkata. [*address*: Kabi Sukanta Cooperative Housing Society Ltd., Flat no. 6, Plot-19 B, CIT Scheme VIII M, Kolkata - 700 067, West Bengal, *e-mail*: <r1938s@gmail.com>].

Bhu Dev Sharma, founder president of WAVES, former president of Hindu University of America and worked as mathematics professor in USA, West Indies, and India. Edited three volumes on Indian Studies, published over hundred articles on mathematics and statistics, guided 23 Ph.D. students, and is Editor of several journals [*address*: Professor of Mathematics, JIIT University, Noida - 201 307 (UP); *e-mail*: <bhudev_Sharma@yahoo.com>]

Bal Ram Singh is the Director of Center for Indic Studies at UMass Dartmouth, where he teaches a course on Science of Kriyayoga. Dr Singh as a Professor of Biophysical Chemistry and Henry Dreyfus Teacher-Scholar at University of Massachusetts Dartmouth, and the Director of Botulinum Research Center, has been conducting research for 20 years at UMass Dartmouth on the molecular mode of action of botulinum and tetanus neurotoxins, and lately also on yoga, mind, and consciousness. His research on biodefence and biotechnology is internationally recognised, and he has served on many national and international scientific panels. Dr Singh has published about

175 research articles; three dozen scholarly articles on issues related to Indian tradition, culture, and philosophy; has edited three books, including *India's Intellectual Traditions and Contributions to the World* (DKPW, 2010); and has obtained 3 patents. Dr Singh has been honoured with awards by several reputed institutions in India and abroad [*address*: Director, Center for Indic Studies, University of Massachusetts, Dartmouth, 285 Old Westport Road, Dartmouth, MA 02747 (USA); *e-mail*: <bsingh@umassd.edu>].

R.P. Singh is chairperson, Centre for Philosophy, School of Social Sciences, Jawaharlal Nehru University, New Delhi - 110 067. He has been doing teaching and research wrote over the last 27 years. His areas of specialization include Modern Western Philosophy, Postmodern Philosophy, Indian Philosophy specially Upaniṣads, Buddhism and Advaita Vedānta. He has to his credit 16 books and 125 articles published in India and abroad. He has delivered lectures and participated in seminars at the Centre for Indic Studies, University of Massachusetts, Dartmouth, USA, University of West Indies, Trinidad and Tobago and many other universities in Europe. He has been thrice awarded *Man of the Year* 2001, 2004 and 2006, by American Biographical Institute. He has been included in *Marquis Who's Who in the World 2009* for valuable contribution in the field of education. Also he has been honoured as one of the 100 *Outstanding Intellectuals of the 21st Century* by the International Biographical Centre, Cambridge University, London, UK, 2010 edition. He won the *International Plato Award*, of the International Biographical Centre, Cambridge University, London, UK, 2011 edition [*address*: Centre for Philosophy, School of Social Sciences, Jawaharlal Nehru University, New Delhi - 110 067; *e-mail:* <rpsinghjnu@yahoo.com>].

M.K. Sridhar, Ph.D, is a professor of Sanskrit at the National Degree College (Autonomous), Bangalore. He was a Fulbright Scholar (2000) with South Carolina University, USA. He has presented papers in 35 international conferences, 33 national conferences and has given invited lectures at Harvard, Berkeley, Stony Brook Universities, USA, Oxford University, UK, and several other important universities in India. His publications comprise fifteen books, two monographs and eleven articles in international peer-reviewed volumes. He was the Chief-editor of 'Kannada-English Concise Dictionary (2004)'. He is a recipient of several awards such as Ramakrishna International Sanskrit Award, Ideal Teacher Award, Banglore (2007), Kapali Shastri Veda Puraskar (2000), etc. [*address*: Associate Professor of Sanskrit, National Degree College, Vani Vilsa Road, Basavanagudi, Bangalore - 560 004; *e-mail*: <drmksridhar@gmail.com>]

Shashi Tiwari is engaged in Sanskrit Studies and Vedic research for the last forty-two years. She has participated in several international conferences in India and abroad. Being the General Secretary of WAVES, India, she has herself organised fourteen international conferences on the themes related to the Vedas and ancient literary heritage. She has twenty-one published books to her credit, apart from her contribution of more than hundred research papers to reputed journals and edited volumes. *Ṛgvedīya Aprisūkta, Sūryadevata, Ṛgvedic Studies, Vedic Studies, Glimpses of Vedic and Ancient Indian Civilization, Sanskrit-Lokokti-Kośa*, and *Sanskrit Sāhitya kā Itihāsa* are her awarded publications. Dr Tiwari has been honoured with eleven academic awards by various Academies of India including the prestigious "Sanskrit Sahitya Seva Sammana", 2006 of Delhi Sanskrit Academy for her contributions to Vedic and Sanskrit Studies. About fifteen research scholars have been awarded M.Phil./Ph.D. degrees under her supervision. Dr Shashi Tiwari, presently working in the department of Sanskrit, Maitreyi College, University of Delhi, has also worked in the Hindu University of America, Orlando, USA as a Visiting Professor of Basic Hindu Scriptures [*address*: 54 Saakshara Apts., A-3 Paschim Vihar, New Delhi 110 063; *e-mail*: <shashit_98@yahoo.com>].

Yamini Bhushan Tripathi, is currently professor Medicinal Chemistry Department, at Banaras Hindu University, Varanasi. With a Ph.D in Bio-chemistry Dr. Yamini has over 30 years of teaching experience. He has successfully executed six scientific projects of the Government of India and is working on two more prestigious projects. He has to his credit 98 published papers in the national and international journals, 3 patents and 2 edited books on Ayurvedic medication. Besides organising dozens of seminars, conferences, he has also delivered 24 lectures abroad, mostly in the Western countries. Between 1985–2005 he was conferred with 10 important awards and currently holds the membership of nine academic societies [*address*: Professor, Deparment of Medicinal Chemistry, Institute of Medical Science, Banaras Hindu University, Varanasi - 221 005, *e-mail*: <yaminiok@yahoo.com> <yamini@bhu.ac.in>].

Index